MW00980387

Scour at marine structures

A manual for practical applications

Richard Whitehouse

Published by Thomas Telford Publications, Thomas Telford Ltd,
1 Heron Quay, London E14 4JD

URL: http://www.t-telford.co.uk

First published 1998

Distributors for Thomas Telford books are
USA: ASCE Press, 1801 Alexander Bell Drive, Reston, VA 20191-4400
Japan: Mauruzen Co. Ltd, Book Department, 3–10 Nihonbashi 2-chome, Chuo-ku, Tokyo 103
Australia: DA Books and Journals, 648 Whitehorse Road, Mitcham 3132, Victoria

Cover picture courtesy of HR Wallingford Ltd

A catalogue record for this book is available from the British Library

ISBN 0 7277 2655 2

Typeset by Gray Publishing, Tunbridge Wells, Kent, UK
Printed and bound in Great Britain by Redwood Books, Trowbridge, Wiltshire

To my family

Preface

This book builds directly on an earlier report of the same title produced within the HR Wallingford strategic research series (report SR417). At the start of the research project there was already a vast amount of literature on the topic of scour, but it was distributed among many diverse sources and most of the information was not available in an easily accessible format, especially for application to the marine environment. The earlier volume by Herbich *et al.* (1984) was available, although out of print, and the river engineering monograph on scour by Bresuers and Raudkivi (1991) had just appeared. This lack of readily accessible information for application to the marine environment led to the wrting of the report and subsequently this book. As so often is the case, research proceeds in parallel at different institutes and an updated version of Breusers and Raudkivi with applications to the marine environment, by Hoffmans and Verheij, appeared during 1997. The topics covered by Hoffmans and Verheij and the material presented in this book form an invaluable and complementary source of information on scouring.

The intention with this book has been to provide the reader with a summary of the information relevant to a particular topic and to present conclusions based upon an assessment of the available information. Methodologies for tackling scour problems are also presented. In this way I hope the book is useful to those readers who are dealing with practical problems as well as to those who are investigating the fundamental processes causing scour.

Richard Whitehouse *obtained a degree in physical geography and a doctorate on sediment transport in marine and aeolian environments. He joined Hydraulics Research Wallingford in 1988 where he has worked on consultancy projects concerning sediment transport, with particular emphasis on scour. He is currently based in the Marine Sediments Group where he leads a section working on sediments based research and consultancy projects.*

Acknowledgements

Valuable assistance and comments were received from a number of colleagues during the research and preparation of this book. In particular, I would like to thank Mr J. S. Damgaard and Mr R. L. Soulsby who provided relevant information for the chapters on the wave-current climate and the scour prediction methods at pipelines and piles. Dr B. Brørs (SINTEF, Civil and Environmental Engineering) provided valuable assistance with the review on computational modelling of local scour and developed the pipeline scour model referred to in this book during an EC-funded sabbatical visit to HR Wallingford from Trondheim. Mrs J. Clarkson and Mrs R. Smith helped admirably by typing the many early versions of the manuscript.

This book is published on behalf of the Department of the Environment, Transport and Regions who part-funded the preparation of the earlier report. The development of ideas in the book benefited from some parallel research carried out by the author on a project funded by the Health and Safety Executive (Offshore Safety Division). Helpful discussions were held with Dr A. Ma (SLP Engineering) and his colleagues at this time. Some final editing of the manuscript was completed during the SCARCOST (Scour Around Coastal Structures) project with partial funding from the Commission of the European Communities, Directorate-General XII for Science, Research and Development, under the Program Marine Science and Technology (MAST III) Contract No. MAS3-CT97-0097. The views and information presented in the book are those of HR Wallingford and not necessarily those of the funding agencies.

HR Wallingford is an independent specialist research, consultancy, software and training organisation that has been serving the water and civil engineering industries worldwide for over 50 years in more than 60 countries. We aim to provide appropriate solutions for engineers and managers working in:

- water resources
- irrigation
- groundwater
- urban drainage
- rivers
- tidal waters
- ports and harbour
- coastal waters
- offshore.

Address: Howbery Park, Wallingford, Oxon, OX10 8BA, UK
Internet: http://www.hrwallingford.co.uk

Notation

$A,\ B$	dimensionless coefficients in time-scale equation (5a)
$A = U_w T/(2\pi)$	orbital amplitude of wave motion at the bed
$c_1,\ c_2,\ c_3,\ c_4$	scour depth coefficients
C_D	drag coefficient applicable to depth-averaged current
C_r	resistance coefficient
d	sieve diameter of grains
D	structure length scale across the flow, e.g. pile or pipe diameter or breakwater width
$D_* = d_{50}\left[\dfrac{g(s-1)}{\nu^2}\right]^{1/3}$	dimensionless grain size
d_{cr}	grain diameter which is just immobile for a given flow
d_n	grain diameter for which n% of the grains by mass is finer, e.g. d_{15}, d_{85}
d_0	nozzle diameter of jet
d_{50}	median grain diameter
$d_{50,b}$	median diameter of grains in sea bed
$d_{50,f}$	median diameter of grains in filter layer
e	2·718281828
$\exp(x) = e^x$	exponential function
E	modulus of elasticity
e_{cr}	critical burial depth for pipeline
e_0	initial gap under pipeline
$F_0 = \dfrac{U_0}{[(s-1)gd_{50}]^{\frac{1}{2}}}$	densimetric Froude number
f_v	vortex shedding frequency
f_w	wave friction factor
g	acceleration due to gravity = 9·81 ms^{-2}

h	water depth
H	height of water wave
H_s	significant wave height
h_t	toe water depth at seawall
I	moment of inertia
k	wave number of water waves $(= 2\pi/L)$ or turbulent kinetic energy
$KC = \frac{U_w T_w}{D}$	Keulegan–Carpenter number
k_s	Nikuradse equivalent sand grain roughness
l	cylinder height
L	wavelength of water wave
L_M	mean wavelength of water waves
L_{max}	maximum span length of pipeline
L_p	length of bed protection
L_s	pipe stiffness length
L_v	wake vortex length
L_0	deep-water wavelength
ln	natural logarithm (to base e)
m	flow speed-up multiplier, due to presence of structure, or submerged weight of pipe
M	shear stress amplification factor, due to presence of structure
MWL	mean water level
N	number of waves
p	shape factor in scour development curve or probability
P	cumulative exceedance probability
q	sediment transport rate per unit time per unit width
Q	volumetric discharge rate from jet
q_b	bedload sediment transport rate per unit time per unit width
r	radial distance from central axis of jet
R	radius of cylinder
$Re_p = \frac{UD}{\nu}$	pile Reynolds number
$s = \rho_s/\rho$	ratio of densities of grain and water
sin	sine
S	scour depth
SWL	still water level

S_e	equilibrium scour depth
S_{max}	maximum scour depth
S_s	settlement depth
$\mathrm{St} = \dfrac{f_v D}{U}$	Strouhal number
S_{3000}	scour depth after 3000 waves
t	time
tan	tangent
tanh	hyperbolic tangent
T	characteristic time-scale for scour or period of water wave
T_p	period at the peak of wave spectrum
T_w	period associated with the amplitude w of the wave bottom orbital velocity
T_z	zero-crossing period of water waves
T^*	dimensionless time-scale for scour
t_1	duration of first time step
u	instantaneous fluid velocity
U	horizontal component of water velocity
$\bar{U}$	depth-averaged current speed
U_c	current speed
U_{cr}	threshold current speed for motion of sediment
$\bar{U}_{cr}$	threshold depth-averaged current speed
U_{rms}	root-mean-square wave orbital velocity at sea bed
U_w	wave orbital velocity amplitude at sea bed
U_{wcr}	threshold wave orbital velocity
$u_* = (\tau_0/\rho)^{1/2}$	friction velocity
u_{*cr}	threshold friction velocity for motion of sediment
U_0	jet velocity
W	scour hole width
w_s	settling velocity of isolated sediment grains
x	horizonal coordinate or distance downstream from centre of pipe
x_s	lateral extent of scour pit from cylinder wall
x_{sw}	distance from sea wall
x_w	radial distance from cylinder wall
y	horizontal coordinate orthogonal to x
z	height above sea bed
z_n	height of jet nozzle above bed

z_0	bed roughness length
δ	boundary-layer thickness
ϵ	porosity of bed or rate of dissipation of turbulent kinetic energy
ζ	bed level, relative to an arbitrary datum
$\theta = \dfrac{\tau_0}{g(\rho_s - \rho)d}$	Shields parameter
θ_{cr}	threshold Shields parameter
$\theta_{2.5}$	Shields parameter calculated with $k_s = 2{\cdot}5d_{50}$
θ_∞	ambient value of θ away from the structure
μ	molecular viscosity or coefficient in Equation (48)
$\nu = \mu/\rho$	kinematic viscosity of water
π	3·141592654
ρ	density of water
ρ_B	bulk density of bed
ρ_s	density of sediment grains
τ_c	current-only shear-stress
τ_{cr}	threshold bed shear-stress for motion of sediment
τ_{jet}	wall shear stress produced by an impinging jet
τ_{max}	maximum bed shear-stress during a wave cycle under combined waves and currents
τ_w	amplitude of oscillatory bed shear-stress due to waves
$\tau_{\beta cr}$	threshold bed shear-stress on a bed sloping at angle β to the horizontal
τ_0	bed shear-stress
τ_{0f}	form drag component of τ_0, due to bedforms
τ_{0s}	skin-friction component of bed shear-stress
ϕ	angle between current direction and direction of wave travel
ϕ_i	angle of repose of sediment
ϕ_c	direction of current
ϕ_w	direction of waves
$\omega = 2\pi/T$	(absolute) radian frequency of waves

Contents

Illustrations

Tables

Figures

Introduction

1. Introduction

Marine structures, whether permanent installations or temporary construction works, are vulnerable to the erosion of the sediment around their base due to the scouring action of waves and tidal currents. The scour pits thus formed can jeopardise the integrity of the structure and must be accounted for at the design stage. Therefore, the first stage of the design process must include an assessment of the likelihood for significant scour to occur and, if necessary, either how the design can be modified to reduce the risk of damage arising from scour or, alternatively, what preventive measures can be taken to reduce the occurrence or to nullify the effects of scour. It is important to design for scour and include prevention measures from the start of a project as the remedial works can be expensive. The aim of this book is to provide methods and guidelines for the engineer to adopt when assessing sediment mobility in the vicinity of sea bed founded structures, in order to be able to gauge the effects of scour on the performance of the foundations and on the structure itself. It should be noted that deposition of sediment at structures could be as significant to their performance as scour but the techniques described here are predominantly for the scour (bed lowering) situation.

The need to address the potential for scour around offshore installations is one of the good engineering practices encouraged in guidelines for the design, construction and certification of offshore installations (Department of Energy/Health and Safety Executive, 1992/93). The influence of toe scour as one of the potential damage sources for coastal structures is recognised (Oumeraci, 1994a). Scouring can adversely affect the stability of the structure and as a result engineers must design for local scour problems as far as possible and monitor and manage any ongoing scour problem in the field. In addition it may be necessary to consider the scouring of soils during the construction phase of a project associated with, for example, the installation of a cofferdam. Table 1 contains a summary of some of the varied types of marine scour problem, and methods of tackling them, using as an example projects undertaken at HR Wallingford.

To be able to design for the occurrence of scour, one needs to be able to predict realistically the maximum depth of the scour hole for the prevailing conditions at a site. Safety factors can be added subsequently, along with the development in space and time (i.e. rate of scour) of the scour pattern. A methodology is required which takes data on the design of the structure, the environmental forcing (waves, currents, etc.) and soil characteristics and converts them into an estimate of the scouring.

In the marine environment the time-varying nature of the waves and currents makes the problem considerably more complex than that of scour at structures in rivers, although the vast scour literature for rivers (e.g. covered by Breusers and Raudkivi, 1991) provides useful background information for the quasi-steady flow situations. Herbich *et al.* (1984), Fowler (1993) and the recent text by Hoffmans and Verheij (1997) have addressed certain aspects of marine scour.

Despite the fact that there are many similarities between all scour problems, and that it is primarily the bed shear stress which causes scour regardless of whether the flow is wave-alone or current-alone, or the combined wave–current case, many different approaches have been adopted. The work described in this book represents an attempt to pull together, into a unified framework, techniques used at HR Wallingford and elsewhere for predicting the depth, extent and rate of scour, in order to facilitate the safe design of structures on sand beds in estuaries and the sea. To enable engineers to make scour predictions in the marine environment, techniques for predicting and combining the long-term frequency distributions of wave orbital velocities, tidal currents and bed shear stresses are provided and the experience from laboratory and field experiments is referred to, as the majority of the methods rely on parameterisation with appropriate data. Where appropriate, the techniques draw on procedures contained in *Dynamics of marine sands* (Soulsby, 1997) and the *Estuarine muds manual* (Delo and Ockenden, 1992).

The book discusses the way in which the presence of the structure disturbs the ambient flow field and modifies the bed

Table 1. Examples of marine scour studies carried out at HR Wallingford

		Environment			
No.	Structure	Waves (W)	Current (C)	W+C	Reversing
1	Pipes (free, fixed)	L			
2	Drilling rig		L		L
3	Outfall	L			
4	Coastal and subsea pipelines	L	N		F
5	Large idealised shapes	LN	LN	LN	L
6	Offshore complex	L	L	L	
7	Offshore complex, pipelines, SSCV		LTN		
8	Heavy prism (Free)		L		
9	Jack-up (Free) and jacket legs		LT		LT
10	Heavy cylinder (Free)	LTNF	LTNF	LTNF	LTNF
11	Probability of re-exposure of pipeline	T	T	T	
12	Likelihood of blockage of outfall by sandwaves	T	T		
13	South coast/Channel sea bed mobility			LTNF	
14	Southern North Sea pipeline abandonment			T	T
15	Platform abandonment, sea bed stability and scour	T	T	T	
16	Sea walls and break-waters	LNF		LNF	

Note: L = laboratory tests, T = theory, N = numerical model, F = field tests, Free = free settling, SSCV = semi-submersible crane vessel.

shear stress (Chapter 2). It examines factors relating to the investigation of scour through physical modelling (Chapter 3) and with computational methods (Chapter 4). A framework is presented in Chapter 5 for making an assessment of the sediment mobility and scour potential under design conditions or for a climate of waves and currents. Techniques for preventing or mitigating the effects of scour are discussed in Chapter 6, and Chapter 7 presents case studies relating to the knowledge and prediction of scour at specific marine structures. Two appendices deal with the computational modelling of pipeline scour (based on Brørs, 1997) and predictive equations for sediment and flow parameters (based largely on Soulsby, 1997).

General principles of scour

2. General principles of scour

Scour occurs where the sediment is eroded from an area of the sea bed in response to the forcing by waves and currents. At a structure the lowering of the bed from some previously obtained equilibrium level occurs due to a localised divergence in the sediment transport rate. Bed lowering can also occur due to the migration or change in shape of bedforms.

Scour is most commonly classified as follows:

- local scour – e.g. steep-sided scour pits around single piles
- global or dishpan scour – shallow wide depressions under and around individual installations
- overall sea bed movement – erosion, deposition, bedform movement.

The combined local–global scour development is illustrated in Figure 1 taken from Angus and Moore (1982).

This Chapter deals primarily with the scour that results from a disturbance to the local flow field caused by the placement of a structure upon the sea bed or at the coastline. The presence of the structure leads to an increase both in the speed of the flow in the vicinity of the structure, due to continuity, and in the turbulent intensity of the flow, due to the generation of vortices (coherent parcels of rotating fluid) from the structure. Even in cases where the flow speed upstream of the structure is below the value of the threshold of motion for sediment movement, the flow speed-up adjacent to the structure can amplify the local value of the bed shear stress to levels which exceed the threshold for sediment motion. This means that in some cases, e.g. during neap tides, the sediment is immobile everywhere on the bed except in the vicinity of the structure. Scour is described as *clearwater* scour when the upstream flow is below threshold, i.e. no sediment transport. The structure acts to initiate sediment

Figure 1. Representation of global and local scour development around a jacket structure (reproduced from Angus and Moore, 1982 by permission of the Offshore Technology Conference)

removal from around its base and with time a net deficit of bed material (depression) develops around the structure in the form of a scour pit. Hence scour should be considered by the engineer who is dealing with the placement of a structure upon a sea bed composed of mobile sediment because the very presence of the structure may induce some localised scouring of the sediments.

2.1. OBSERVATIONS OF SCOUR

An industry report (ICE, 1985) recognised the importance of scour for coastal structures and called for more research on scour in support of the civil engineering community. In 1986 a comprehensive report on sea wall performance in the UK was published by CIRIA (1986). The report concluded that toe scour represented the most serious and prevalent cause of damage to sea walls, directly accounting for over 12% of the case histories of sea wall collapse/breaching or loss of fill studied. The problems presented by scour for nearshore and coastal structures and an indication as to the economics of remedial work resulting from settlement and scour in the USA are contained in Hales (1980a) and Lillycrop and Hughes (1993). Hales (1980b) reports that in the USA '... foundation scour has significantly affected the economics and service life of many coastal structures'. Reports of the failure of crucial engineering structures, although of interest to the engineer and scientist working on scour, are usually to the detriment of the company responsible for the structure and are therefore, unless they result in the loss of life, have significant economic impact or are viewed publicly, treated with confidentiality. However, the publication of such information can help to prevent future costly mistakes.

In the case of offshore structures, scour reduces the lateral support for the piles which leads to an increase in bending stresses unless remedial action is taken (Watson, 1979). Scour to depths of between 1·5 and 3·5 m can take place rapidly in the absence of any bed protection. This scouring can produce large depressions beneath structures which both affect its stability, by reducing the effective depth of pile penetration, and expose suspended risers to hydrodynamic loading that may exceed design limits. The occurrence of scour offshore also causes operational difficulties even for short stay jack-up rigs (Sweeney *et al.*, 1988; McCarron and Broussard, 1992).

Watson described some of the experiences of the Amoco (UK) Exploration Company relating to offshore scour. He reported on two different types of scour that had been observed around structures placed upon sandy sediments in the gas fields of the Southern North Sea. The first was *local* scour which produced a pit with the shape of an inverted frustum (truncated cone) located around the centre of the individual legs of structures. In some cases the individual scour pits coalesced to such an extent that they formed a depression over a much bigger area, of the same order as the area of the structure supported by the piles. This coalescence of individual scour pits may have been the source of the second phenomenon known as *dishpan* or *global* scour, in which a depression was observed to form under the whole area beneath an offshore platform or rig (Figure 1). The shape of the dishpan scour pit is an elliptical bowl with its centre based beneath the structure and with the major axis of the ellipse perpendicular to the predominant direction of the current flowing around the structure. The dishpan scour pattern has been observed both in the North Sea (Van Dijk, 1980) and in the Gulf of Mexico (Sybert, 1963), and similar patterns can exist

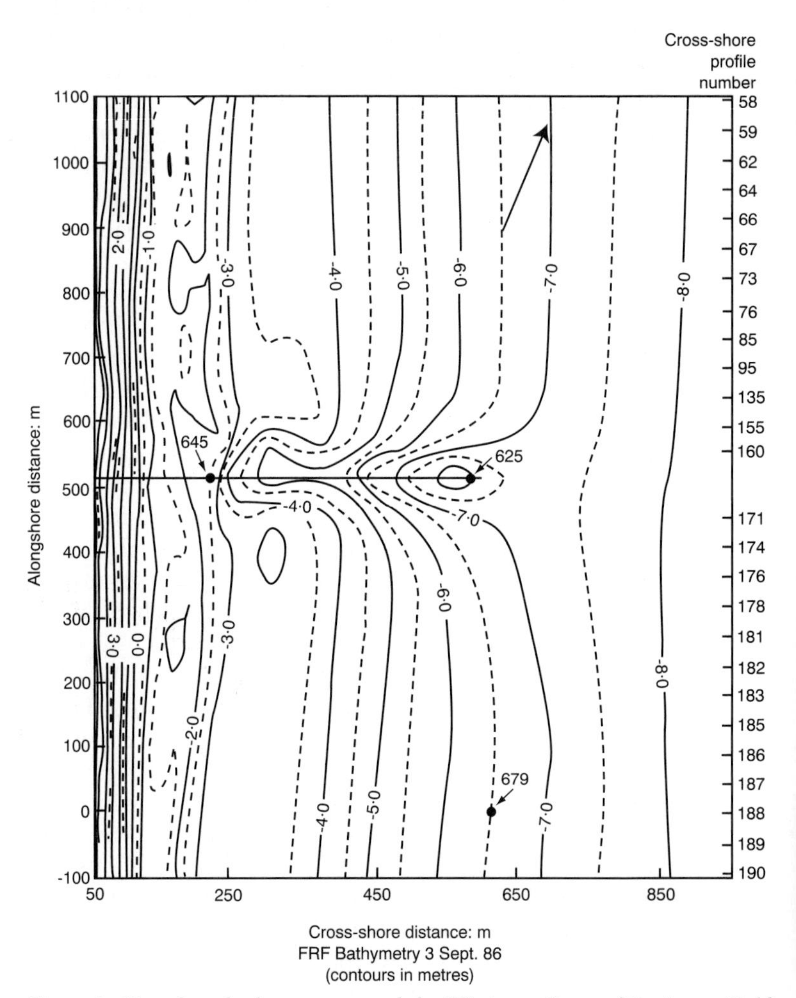

Figure 2. Nearshore bathymetry around the US Army Corps of Engineers Field Research Facility, a piled jetty structure at Duck, North Carolina (reproduced from US Army Corps of Engineers, 1986, by permission of the Coastal Engineering Research Centre, USACE)

under coastal structures, even where these have been designed specifically to have a minimal influence on the nearshore bathymetry (Figure 2). Whitehouse *et al.* (1997) present field data from a pier on the south coast of the UK which suggests the pier can locally modify the hydrodynamics under certain conditions. The creation and evolution of each type of scour hole may be heavily influenced by the interaction of the flow field with the two types, where they both exist. The *dishpan* scour may be caused by more than the coalescence of local scour pits due to a general lowering of the sea bed resulting from the interaction of the wave and current field with the whole structure, rather than individual legs and piles. The mechanisms responsible for its formation remain unclear as it has not generally been possible to reproduce it in laboratory investigations, except in the tests by Posey (1970). Bed liquefaction is most often invoked as the cause of *dishpan* or *global* scour (e.g. O'Connor and Clarke, 1986).

Pipelines are periodically covered and uncovered in areas of mobile sediment (Langhorne, 1980). Exposed pipelines are susceptible to scour and the development of free-spans. Herbich (1985) stated that '... many reported and unreported pipeline failures have occurred since the placement of pipelines on the ocean floor'.

As no broad change of engineering practice has occurred over the past 30 years, it can be concluded that scour remains a problem with structural, economic and safety implications.

2.2. PHYSICAL CONSIDERATIONS

Over the past 30 years a vast number of laboratory investigations of local scour around structures have been reported in the open literature, especially relating to river bridge engineering (e.g. the monograph by Breusers and Raudkivi, 1991).

Two types of local scour are traditionally identified in the river engineering literature. These can be defined in terms of the bed shear stress τ_0, which is the frictional force exerted by the flow on the bed per unit area of bed (Nm^{-2} in SI units), and the corresponding threshold value for sediment motion τ_{cr}.

- *Clear-water* scour where the ambient bed shear stress τ_0 is less than the critical value for sediment motion τ_{cr} but is greater than τ_{cr}/M in the vicinity of the pile, where M is the shear stress amplification factor adjacent to the structure (see Section 2.5.1). In this situation sediment is transported away from the periphery of the pile but is immobile elsewhere on the bed. The equilibrium scour depth is reached when the agitating force due to the flow balances the resistive force of the particles.
- *Live-bed* scour where $\tau_0 > \tau_{cr}$ everywhere on the bed and sediment is being transported by the flow from upstream, through and out of the scour hole. An equilibrium scour depth is also obtained even though the shear stress in the scour hole exceeds the threshold.

The variation of the equilibrium scour depth with hydraulic parameters and the corresponding time taken to reach equilibrium need to be known to enable reliable predictions of the scour development at a structure.

For a full analysis, the stability of the soils and the fluctuations in general bed level not related to the presence of an installation must also be taken into account (including bedforms). The likely magnitude of general bed level changes can be assessed from a historical analysis of bathymetric data or from computational models designed to predict the sea bed evolution over large areas (kilometres) (e.g. Wallace and Chesher, 1994).

2.2.1. *Basic sediment budget equation*

The morphodynamic response of the sea bed, scour or accretion around a structure can be considered in terms of the sediment budget equation

$$\frac{\partial \zeta}{\partial t} = -\frac{1}{1-\varepsilon}\left(\frac{\partial q_s}{\partial x} + \nabla_D - \nabla_E\right) \qquad (1)$$

which, for a one-dimensional (single point) application, defines the change in elevation ζ through time t of a unit area of the bed as governed by the flux of sediment q passing through the unit area plus a term representing the rate of sediment deposition

(∇_D) and another the rate of sediment entrainment into the flow (∇_E). The factor outside the brackets containing bed porosity ε is necessary to convert the volume flux of sediment to the volume of material including pores resting on the bed. The term representing storage of sediment on the sea bed ($\nabla_E - \nabla_D$) is likely to be significant because of the locally rapid spatial variation in q around the structure. Where the structure has a complex shape this term is presently best assessed from physical model investigations.

In two horizontal dimensions (x, y) Equation (1) can also be applied to denote the change in bed elevation ζ at a single location

$$\frac{\partial \zeta}{\partial t} = -\frac{1}{1-\varepsilon}\left[\left(\frac{\partial q}{\partial x}+\frac{\partial q}{\partial y}+\nabla_D(x,y) - \nabla_E(x,y)\right)\right] \quad (2)$$

Methods for calculating q are given by Soulsby (1997).

2.2.2. *Scour depth*

The clear-water scour depth at a pile varies with τ_0 for a set combination of the pile diameter and sediment size but is more or less independent of the value of τ_0 for the live-bed condition (Figure 3). The equilibrium clear-water scour depth S_e reaches a maximum at $\tau_0 = \tau_{cr}$; beyond this point the equilibrium scour depth remains more or less constant but the rate at which scouring takes place increases with τ_0. For practical reasons the two regimes are often treated separately but in reality they are part of a continuum.

2.2.3. *Time-scaling*

Predictive formulae rely on being able to include the effect of the time development of scour. The rate at which the scouring of sediment takes place from around a vertical pile is related to the divergence in the sediment transport rate in the control volume around the pile, which arises from the disturbance to the flow field.

For a given set of environmental conditions the scouring of sediment at structures initially occurs rapidly but then

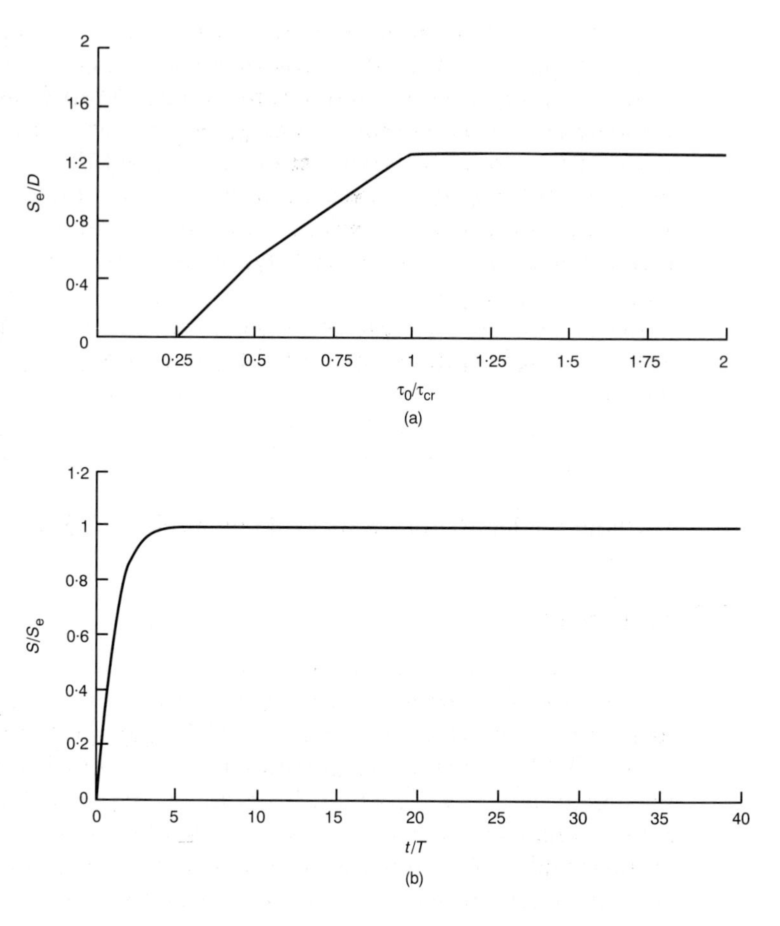

Figure 3. Scour development at a single pile in steady flow: (a) variation of equilibrium scour depth with shear stress; (b) typical time development curve

approaches its maximum value asymptotically. Data for the time development of scour at single piles by a steady flow have been reported by Breusers (1972), Carstens (1966), Chabert and Engeldinger (1956), Franzetti *et al.* (1981, 1982), Kothyari *et al.* (1992), and Yanmaz and Altinbilek (1991) and May and Willoughby (1990, large diameter piles) have both studied square and cylindrical cross-sections. Data for scour development in waves are presented by Sumer *et al.* (1992b) and Kawata and Tsuchiya (1988). Measurements of the increase in depth of the scour pit S with time t at a fixed vertical pile in a steady current and under wave action (Sumer *et al.*, 1992a) have been found to fit well to the following formula:

$$S(t) = S_e\left[1 - \exp\left(-\frac{t}{T}\right)^p\right] \quad (3)$$

where S_e is the equilibrium scour depth that would be attained from a curve-fitting exercise to data as t tends to infinity, T is the characteristic time-scale for scour and p is a fitting coefficient relating to the shape of the $S(t)$ curve, usually taken as being equal to unity (Figure 3*b*). Sumer *et al.* (1992a, 1993) use $p = 1$ both for wave scour and for current scour. The time-scale T is defined as the time after which the scour depth has developed to 63% of the equilibrium value.

Within the limits of experimental scatter, and as far as can be determined from the available data, the value of p appears to be nearly invariant with flow speed for any particular flow-structure geometry and any dependency of p on external variables has not been identified.

The experimental data for the time development of live-bed scour can be collapsed into a single relationship through the dimensionless time-scale T^* obtained at HR Wallingford based upon the following premises:

- the dimensions of the scour-pit scale geometrically with the diameter D of the pile or pipeline (or an appropriate measure of the cross-flow dimension of the structure)
- a generalised form of the sediment transport formula
- the sediment budget equation (Equation 2).

This results in a formula identical to the one published by Sumer *et al.* (1992a) and Fredsøe *et al.* (1992):

$$T^* = T\left[g(s-1)d_{50}^3\right]^{0.5} D^{-2} \quad (4)$$

The previously undefined symbols are g the gravitational constant, s the sediment mineral specific density ρ_s/ρ (normally about 2·65 for quartz sand) and d_{50} the median grain size of the sediment.

The value of T^* is correlated with the non-dimensional bed shear stress as follows:

$$T^* = A\theta_\infty^B \tag{5a}$$

where

$$\theta_\infty = \frac{\tau_0}{(\rho_s - \rho)gd_{50}} \tag{5b}$$

A and B are constants for a given geometry and θ_∞ is related to the ambient flow, i.e. away from the structure. The numerical value of B is found to be negative and thus this relationship demonstrates that the scour takes place faster as the shear stress increases. The use of the non-dimensional time-scale facilitates the application – by similarity of the Shields parameter (Equation (5)) – to other flows, sediments and (cylinder) diameters, provided the value of θ is obtained from the ambient flow speed close to the bed and away from the structure.

The empirical constants in Equation (5a) are determined by a best fit to steady flow data for waves or currents which then allows other non-steady flow scenarios to be modelled as a series of quasi-steady steps (see Section 2.2.4 below).

For cases other than piles and pipelines, which are discussed in Chapter 7, it may be appropriate to determine a statistical mean value of p corresponding to a particular combination of structure geometry and flow conditions from (laboratory) measurements of scour development.

As the prediction method is written in terms of the dimensionless bed shear stress θ, we can potentially include the effect of waves and currents in the scour predictions. Combined wave–current shear stresses can be calculated using the methods outlined in Appendix 2. However, further analysis of existing data will be required, probably in combination with some further laboratory testing, to develop fully the prediction method to work for the combined flow situation.

2.2.4. *Time stepping approach to scour predictions*

In order to predict the scour development in non-steady flow, a time stepping approach has been devised (Gilbert, G., personal communication, 1993). The principle of the time stepping model is that an increment in scour depth δS may be calculated for a time interval δt when a quasi-steady flow is assumed to exist (a reasonable approximation if δt is small enough). Then it is assumed

$$\delta S = \frac{\mathrm{d}S}{\mathrm{d}t}\,\delta t \tag{6}$$

From Equation (3):

$$t = T\left[-\ln\left(\frac{S_e - S}{S_e}\right)\right]^{1/p} \tag{7a}$$

$$\therefore\ 1 = \frac{T}{p}\left[-\ln\left(\frac{S_e - S}{S_e}\right)\right]^{\frac{1}{p}-1}\frac{\mathrm{d}S/\mathrm{d}t}{S_e - S} \tag{7b}$$

$$\text{i.e.}\ \frac{\mathrm{d}S}{\mathrm{d}t} = \frac{p(S_e - S)}{T\left[-\ln\left(\frac{S_e - S}{S_e}\right)\right]^{\frac{1}{p}-1}} \tag{8}$$

where ln is the natural logarithm to base e.

Thus the calculation of the scour depth increment δS during any time step requires knowledge of the length of the time step, δt, the degree of scour which has already occurred S, the ultimate scour depth S_e, the time-scale T and the empirical constant p. The equations for T and S_e are given in the previous section. It will be noticed that an anomaly arises in Equation (8) when $S = 0$ which leads to a denominator of zero since ln $(1) = 0$. The time-stepping model uses the analytical Equation (3) to calculate the scour occurring over the interval $0 \leq t \leq \delta t_1$ where δt_1 is the first time step's length. Subsequent time steps may then always utilise a non-zero value of S. The condition $S > S_e$ will lead to negative scour which is not physically meaningful in our context and so $S > S_e$ must give $\mathrm{d}S/\mathrm{d}t = 0$ within the model.

The length of the time step, δt, will be chosen by considering the other time parameters in order to give adequate resolution in time of the physical processes. It must be much smaller than T, the time-scale of scour, and much smaller than the period of time over which the flow inducing scour is varying. A factor of 1% of the smallest of these time parameters should be sufficient to calculate δt during a simulation. This method has been checked and gives the same result for steady flow as the analytical approach, Equation (3).

Physically the scour development in a non-steady flow will be a combination of excavation and backfill as the magnitude of the bed shear stress changes. Whilst the model described above does not explicitly include the backfilling process, this kind of approach, using steady state wave scour results to simulate data for changing wave conditions, has been verified for wave scour below pipelines by Fredsøe *et al.* (1992).

2.3. FLOW DISTURBANCE AROUND VERTICAL PILES

The sea bed boundary layer flow approaching a vertical cylinder sets up a pressure gradient on the upstream face of the cylinder, between the low pressure in the near-bed flow and the high pressure in the flow above, which drives a flow down the face of the cylinder (the downflow, Figure 4). A recirculating eddy (primary vortex) is formed when the flow impinges with the sea bed, this wraps around the cylinder and trails off downstream

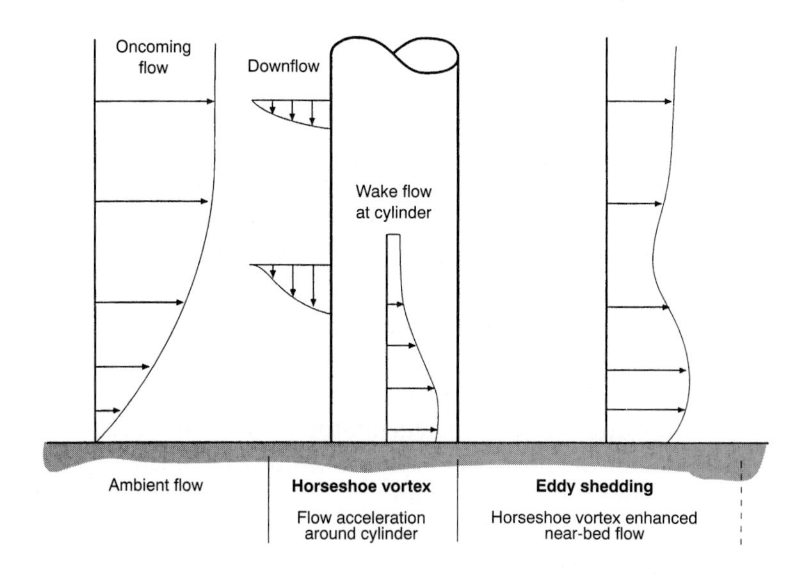

Figure 4. Flow–structure interaction for a vertical cylinder

from points half-way around the cylinder to create the second *horseshoe* shaped vortex. From a sediment transport point of view the primary and *horseshoe* vortices are the major mechanism leading to the scouring of sediment from around the base of a vertical cylinder. The characteristic scour pattern is illustrated in Figure 5. There is some conjecture that the scour hole development actually enhances the strength of the vortex and *downflow* (Breusers *et al.*, 1977).

The characteristics of the vortex system upstream of a pile have been investigated by, for example, Baker (1979) who found that in laminar boundary layer flow the number and position of vortices was dependent upon the pile Reynolds number

$$\mathrm{Re}_p = \frac{UD}{\nu} \tag{9}$$

where U is the steady flow velocity, D the pile diameter and ν the kinematic viscosity of the fluid, and the relative boundary layer thickness δ/D. Shen *et al.* (1966), amongst others, have correlated the scale and intensity of the horseshoe vortex in steady turbulent flow with the pile Reynolds number, a finding which is supported in the data of Hjorth (1975). Bělik (1973), Hjorth (1975) and Baker (1980) have investigated the position of the

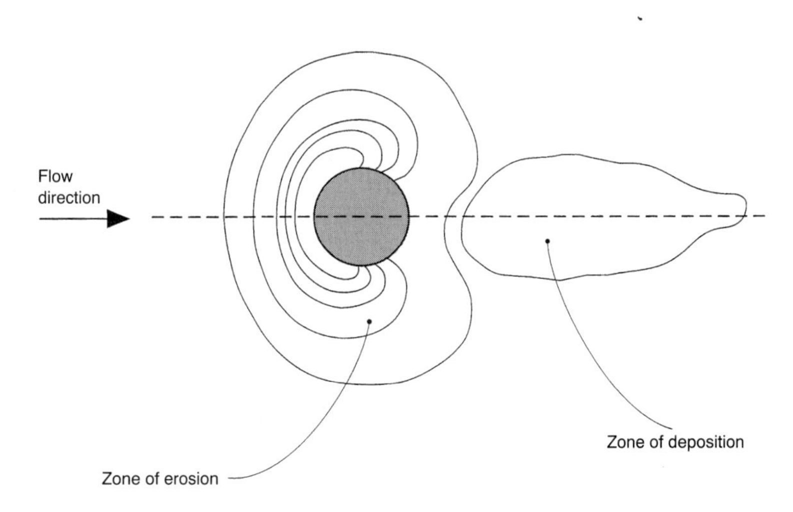

Figure 5. Characteristic equilibrium scour hole pattern for a vertical cylinder in steady flow

separation line and hence the horseshoe vortex upstream of a cylinder in turbulent flow.

For turbulent flow the horseshoe vortex on a flat bed (defined by minima in bed pressure measurements) operates at a distance of between 0·2 and $0{\cdot}3D$ from the upstream face of the cylinder, although there is some indication that it operates closer to the cylinder wall for smaller δ/D than for large δ/D (Bělik, 1973; Baker, 1980). The point of flow separation upstream of the vortex scales with D when $l/D > 1{\cdot}5$ and on l when $l/D < 0{\cdot}3$, where l is the height of the cylinder (Baker, 1980).

Hjorth (1975) investigated in detail the flow pattern around a circular cylinder and found an increasing influence of the pile on the boundary layer flow within a distance of $4D$ from the upstream face of the cylinder. The streamlines in the disturbed region radiate out from the upstream face of the cylinder and are then bent round and realign with the ambient flow direction downstream. Laterally, the velocity within a distance of one cylinder radius from the pile wall is enhanced and the vertical profile is more uniform as the vorticity is concentrated by convection to a thinner layer near to the bed. The nearest unaffected streamline at the bed passes within one cylinder radius from the cylinder wall and thus within this region water is fed into the near bed flow from higher up in the water column. The pattern of streamlines for rectangular cylinders is similar but laterally, within a distance $D/2$ from the wall, the gradient of the vertical velocity profile is negative due to the vortex action near the cylinder. The flow pattern around both types of cylinder was found to be independent of the turbulence intensity in the approach flow.

The *wake vortex* is formed by the rolling up and separation of the unstable shear layers generated around the structure and gives rise to eddies being shed downstream in a periodic fashion. The wake vortex system acts somewhat like a vacuum cleaner in picking up bed material over a downstream distance of about $8D$. The shedding frequency of the vortices occurs at a Strouhal number

$$\mathrm{St} = \frac{f_V D}{U} \tag{10}$$

of approximately 0·2 in the pile Reynolds number range 10^2–10^5. In higher Reynolds number flows, e.g. high current speeds or large cylinders, the periodicity with which the vortices are shed becomes chaotic.

Sumer *et al.* (1992b) have investigated the existence of the horseshoe vortex in oscillatory (wave) flow and concluded that a lower limit to the existence of this vortex was at a value of the Keulegan-Carpenter number equal to 6, i.e

$$\mathrm{KC} = \frac{U_{\mathrm{W}} T_{\mathrm{W}}}{D} \geq 6 \tag{11}$$

where U_{W} and T_{W} are the amplitude of the bottom orbital velocity and associated period and D is the pile diameter. A small value of KC is analogous to only a very thin boundary layer with respect to the diameter of the pile. As the KC number and wave boundary layer thickness increase the horseshoe vortex exists for a larger portion of the wave cycle (up to about 50% of the time). The locations of the primary vortex, trailing horseshoe vortex and eddy shedding are similar to the current alone case.

In the case of the combined action of waves and currents the horseshoe vortex may be less strong but, none the less, still important. If the steady current is much stronger than the peak bed orbital velocity of the wave motion, then the horseshoe vortex due to the steady current will exist but it is periodically enhanced and reduced by the oscillatory motion. At the other extreme, if the peak bed orbital velocity due to the waves is much larger than the ambient current then the horseshoe vortex may exist only in a very weak form, if at all. Little is known about the case with approximately equal oscillatory and steady current flow velocities. It could be that modified mechanisms act in this situation, and the literature currently provides conflicting evidence for both an exacerbation and an alleviation of the scour problem in combined wave–current flow (see Chapter 7).

2.4. FLOW DISTURBANCE AROUND PIPELINES

From the time that a pipeline is placed upon the sea bed the flow passing over the pipeline separates, resulting in an area of recirculating flow being produced downstream of the pipeline to a distance of between $6D$ and $10D$, the point at which the flow streamline reattaches itself to the bed. Eddies are also shed from the top of the pipeline in the form of the *lee wake* (Figure 6, from Sumer and Fredsøe, 1990) which produces a periodic fluctuation in the bed shear stress in the region $1 \leq x/D \leq 8$ from the pipe with values periodically above and below the mean value. The fluctuating shear stress will enhance the potential for scour (Mao, 1986). The significance of the lee wake is increased under the action of waves or a tidal current as the vortices are shed on different sides of the pipe for different phases of the wave/tide cycle. The vortex shedding occurs at a value for the Strouhal number of about 0·22, although possibly higher than this early in the scour development. There are also two other vortices which act under the separation streamline and adjacent to the pipe–soil boundary, a clockwise rotating eddy of diameter $D/2$ upstream of the pipe and an anticlockwise eddy of the same size downstream.

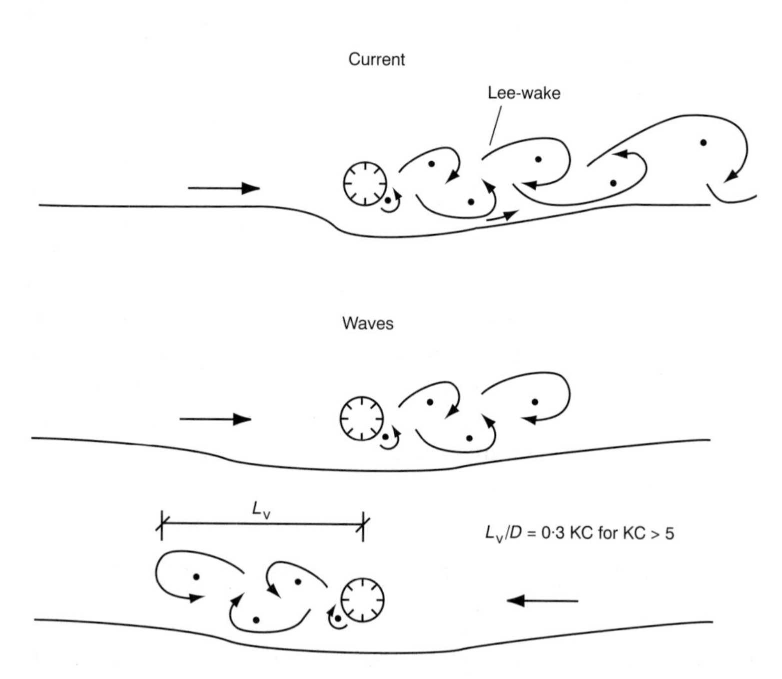

Figure 6. Flow structure interaction for a pipeline (reproduced from Sumer and Fredsøe, 1990, by permission of the ASCE)

The pressure gradient across the soil element under the pipe (positive pressure upstream, negative pressure downstream) is an important factor in initiating scour underneath the pipeline (Hjorth, 1975; Sumer and Fredsøe, 1991). Once a sufficiently strong seepage flow occurs under the pipe then a sudden localised removal of bed material by *piping* occurs, aided by the excavating vortices described above. These may excavate sediment from under the pipe regardless of the water flow through the soil and will speed up the onset of scour. These processes are important in currents (Kjeldsen *et al.*, 1974) and waves (Mao, 1986). The next stage of the erosion process, at which point water starts to pass through the confined space under the pipe, is called *tunnel erosion*. Scouring in this phase occurs very rapidly due to the fourfold amplification in bed shear stress under the pipe in the early stages (Sumer and Fredsøe, 1991, 1993). Tunnel erosion can lead to the complete undermining of a section of pipeline and the generation of a free span. The lee wake erosion acts in conjunction with the tunnel erosion and can lead to the formation of wide scour pits (see Chapter 7). In reality the sea bed will not be flat and the process of tunnel erosion may occur preferentially at a low spot in the sea floor topography under the pipeline.

2.4.1. *Flow around short horizontal tubulars*

Scour proceeds preferentially around the ends of a short cylinder lying on the bed because by convergence of the streamlines the flow speed is enhanced within a distance from the end of the cylinder equal to the height (diameter) of the cylinder. A gradual erosion of the soil bearing area beneath the cylinder takes place from the ends inward and with time the cylinder settles into the bed in a stepwise fashion (Carstens, 1966).

2.5. SCOUR MECHANISMS

The characteristics of the disturbance to the ambient flow depend on the shape and size of the structure and its orientation with respect to the ambient flow direction. The blockage offered to the flow by large structures will be greater than from slender ones, and angular or irregularly shaped structures will produce a more complex and turbulent flow than that formed around streamlined structures.

2.5.1. *Amplification of bed shear stress*

Potential flow solutions for the unseparated flow around a vertical cylindrical pile or horizontal pipeline indicate that the flow speed-up is a factor of 2 which, assuming bed shear stress is proportional to the square of the flow speed (Appendix 2), produces a factor of 4 amplification in the shear stress, i.e. $M = m^2$. However, the shear stress enhancement in complex geometry situations is best obtained from laboratory measurements and examples of published values are given in Table 2 and discussed below.

The amplification of the shear stress around a pile in steady flow has been discussed by Niedoroda and Dalton (1982) who refer to the measurements of Schwind (1962) and Hjorth (1975). These studies showed that the maximum amplification of the

Table 2. Typical values for shear stress amplification M around structures

Flow/structure type	M	Comments/Source
Steady flow		
Slender vertical pile	4	Additionally (Hjorth, 1975) M varies with h/D, can be larger than 4 – see text
Pipeline	4	Value under pipeline after initial breakthrough (Sumer and Fredsøe, 1993)
Squat pile	2 to 5	(Baker, 1979 ; Eckman and Nowell, 1984)
Bridge caisson (rectangular)	4 to 6	(Hebsgaard *et al.*, 1994)
Wave flow		
Circular pile	2 to 3+ (fn: KC no.)	(Sumer *et al.*, 1992b)
Wave–current flow		
Bridge caisson (rectangular)	4 to 6	(Hebsgaard *et al.*, 1994)

mean bed shear stress beneath the horseshoe vortex may be respectively eleven or twelve times greater than the ambient value although Neidoroda and Dalton felt that there was some uncertainty as to the range of validity of these results in the marine case. Baker (1979) measured a fivefold increase in the skin friction coefficient or bed shear stress upstream of a squat pile under the primary horseshoe vortex, for laminar flow, and Eckman and Nowell (1984) have measured a greater than two times amplification within a distance D of the outer wall of a short vertical cylinder.

A closer examination of the data of Hjorth (1975) shows that for a given combination of flow depth and circular cylinder diameter D, the magnitude of the maximum shear stress amplification M appears to vary slightly with ambient flow speed U. There also appears to be an overall increase in M with pile Reynolds number UD/ν, assuming that ν is constant in his experiments. Taking an average of the values of the maximum value of M adjacent to the cylinder (for data with a given water depth h and the flow speeds tested at that depth) indicates that for $h/D < 2$ the value of M adjacent to the cylinder wall is around 10 but for deeper water (larger h/D) M decreases linearly with increasing h/D to a value of 6 at h/D equal to 4. Thus the data are suggestive of M decreasing to the usually accepted value $M = 4$ for a slender circular cylinder in deep water, occurring in Hjorth's data at h/D equal to about 5. This apparent dependency of M on h/D possibly arises due to the variation in the position of the horseshoe vortex with h/D, as h/D decreases the vortex moves closer to the cylinder (Baker, 1980). The patterns of shear stress amplification shown in Figure 7a are similar for all cases tested by Hjorth, i.e. maximum values at a radial location about 45° from the axis of the flow.

In summary:

- the horizontal scale of the amplification appears to be determined by the diameter of the cylinder
- at least in the region between the axis of the oncoming flow and around to the position at which the shed eddies become dominant, the amplified shear stress appears to be confined to a circle (concentric to the cylinder) of radius equal to the diameter of the cylinder.

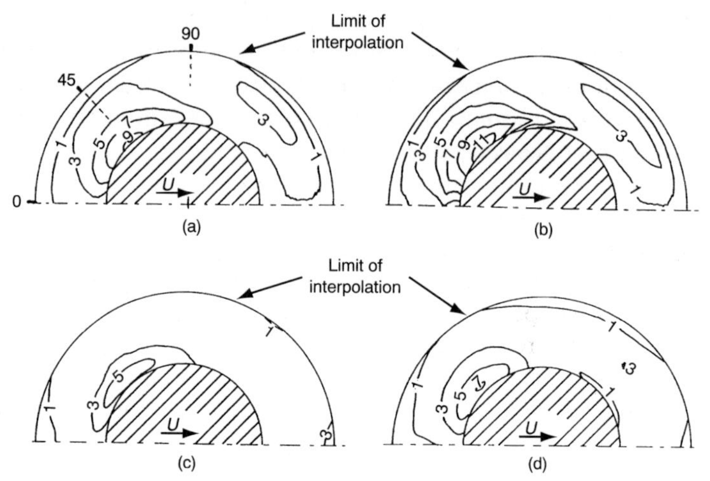

Shear stress distribution around a 5 cm circular cylinder in steady flow. Water depth 10 cm and (a) $U = 15$ cm s^{-1}, (b) $U = 30$ cm s^{-1}. Water depth 20 cm and (c) $U = 15$ cm s^{-1}, (d) $U = 30$ cm s^{-1}

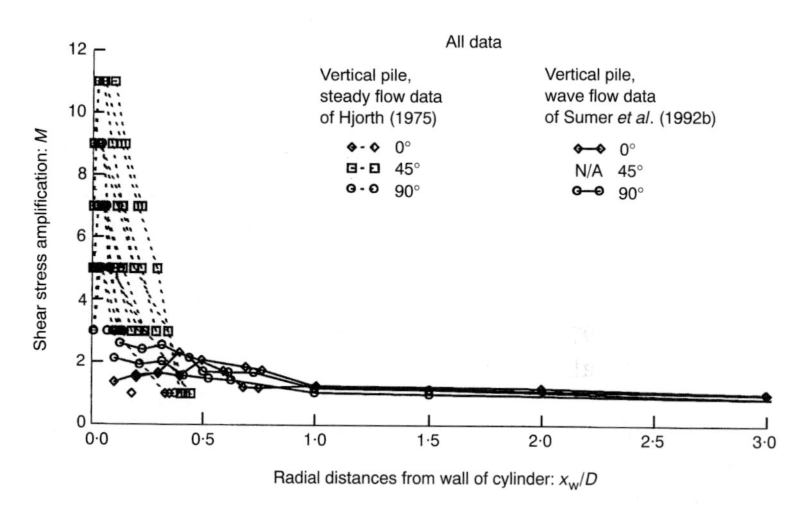

Figure 7. Variation in shear stress amplification away from the wall of a vertical cylinder (upper panel after Hjorth, 1975, reproduced with permission of the author)

The amplification in shear stress has been deduced for other cases as follows.

Sumer *et al.* (1992b) published direct measurements of the bed shear stress around a pile in oscillatory wave flow. Their results, which are shown in Figure 7b, indicate an amplification in the bed shear stress adjacent to the cylinder by a factor of between 2 and 3, over a streamwise and transverse distance of about $1{\cdot}5D$, but with dependence upon the KC number.

For the case of a rectangular cylinder with its flat face pointing into the flow the value of M is only about 3, whereas M is at least a factor of three times larger than this when the corner points into the flow (Hjorth, 1975). Thus the pattern and magnitude of M for this shape of structure is highly dependent on the orientation with respect to the flow.

O'Riordan and Clare (1990) performed numerical computations of the velocity amplification around a square gravity base structure for two cases, flow towards a corner and flow towards a face. The velocity (shear stress) was amplified by factors of 3·2 (or 10·2) and 2·0 (or 4·0) respectively. The larger amplification for the flow towards the corner is in accord with the findings of Hjorth (1975).

Hebsgaard *et al.* (1994) measured in small-scale model tests the shear stress amplification around large (30 m by 17 m prototype) bridge caissons in 30 m (prototype) of water. They obtained values of M between 4 and 6 at 1 m from the pier in steady currents and waves and current flow.

For pipelines, once the scour hole has begun to develop under the pipe the initial value of M is 4 (Sumer and Fredsøe, 1993). For a gap under the pipe of $0{\cdot}25D$ the value of M is 3·1 (Mao, 1986). Hjorth (1975) measured the amplification of shear stress around a pipe half-buried in a flat bed and found that M was generally about 1 (no amplification) on the flat bed either side of the pipe, although the shear stress fluctuation was generally higher than for the case with no pipe.

In summary:

- flow speed up around the structure results in locally higher values of bed shear stress
- this effect can be parameterised with the factor

$$M = \tau_0(\text{local})/\tau_0(\text{ambient})$$

- the maximum value of M occurs adjacent to the structure
- M becomes smaller with increasing distance from the structure
- the distribution of M around the structure directly controls the pattern of scouring
- in the marine environment the varying angle of wave–current attack produces a time variation in the location (and pattern) of M.

The variability in the reported values is probably partly due to differences in the thickness δ of the bottom boundary layer (or water depth h) with respect to the principal diameter of the structure. The differences between the steady flow and wave flow situation are (1) the prototype tidal boundary layer is much thicker (at least several metres) than the one associated with wave action (typically several centimetres) and (2) the fact that the horseshoe vortex does not exist for the whole of a wave cycle (Sumer *et al.*, 1992b) whereas it is persistent in a steady flow.

2.5.2. *Bed liquefaction due to waves*

An important mechanism which makes the bed surface layers more susceptible to erosion is bed liquefaction (e.g. Sakai *et al.*, 1992). Steep storm waves are likely to be the most effective at causing the liquefaction of the bed because they generate high pressure gradients at the bed. Liquefaction has been suggested as one possible cause of the global scour phenomenon (e.g. O'Connor and Clarke, 1986). Liquefaction is also considered important for estimating scour at, and hence the stability of, coastal structures (e.g. Zen and Yamakazi, 1993).

A bed is in a liquefied state when it has very low or zero shear strength, i.e. the grains within the bed are unconstrained by neighbouring grains. This has two effects: (1) it removes the capacity of the bed to support a normal load (zero bearing capacity) and (2) it makes the bed much more susceptible to erosion by waves and currents because of the reduced intergranular friction. Watson (1979) has commented that the presence of a structure on the bed (i.e. confining pressure) may make the bed sediment more susceptible to liquefaction.

Sakai *et al.* (1992) have discussed the mechanisms by which liquefaction can be brought about. Surface waves induce a hydrostatic pressure cycle at the bed surface, in which a high pressure accompanies the passing of a wave crest and a low pressure accompanies the passing of a trough. These pressures are transmitted into the bed and they give rise to horizontal and vertical pressure gradients which encourage liquefaction behaviour (*oscillatory excess pore pressure*, Zen and Yamakazi, 1993). The method of Sakai *et al.* (1992, poro-elastic sea bed) can be used to predict times within the wave cycle when bed liquefaction is likely to occur, although the predictions are extremely sensitive to the degree of aeration of the sediment pore water which exerts a strong control on the effective bulk modulus of the pore water. The pressure cycle generated by surface waves may also bring about a gradual increase in the interstitial pore pressure in conditions where the excess pressure accompanying a wave crest cannot be fully dissipated. In some cases this pore pressure becomes so great that it loosens the bed grains to the point where they no longer experience a significant force from neighbouring grains and liquefaction occurs (*residual excess pore pressure*, Zen and Yamakazi, 1993). This longer-term pore pressure build-up is also known as *cyclic loading*.

2.5.3. *Other mechanisms promoting scour*

The rocking of structures by waves in storms can also lead to a build-up in soil pore water pressure beneath the foundations until the pore water escapes through weaker channels in the fabric of the sediment, known as *piping*. Piping erosion of the sea bed, due to the differential pore water pressure which occurs under foundation loading and wave action, was postulated (Bishop, 1980) as the cause of failure of the foundation of a research tower placed in Christchurch Bay off the south coast of England. The conventional wave–current scour was exacerbated by the cyclic rocking of the tower.

The cyclic loading and associated rocking of piles by wave action can produce a localised scouring of sediment, as has been studied experimentally by Reese *et al.* (1989). Onshore tests in cohesive soil showed that loss of lateral capacity can occur due principally to the creation of a gap at the pile–soil interface both mechanically and by the ejection of water from the gap as it opens and closes.

At certain locations the potential for fish and other biota to excavate sediment from around structures might also need to be considered.

Physical modelling of scour

3. *Physical modelling of scour*

3.1. DESCRIPTION OF PHYSICAL MODELS

Many scour problems are investigated by the use of physical modelling using a tank in which the wave and current conditions can be simulated with a scale model of the structure placed upon a mobile bed of (scaled) sediment. Many studies are concerned with the scour that occurs on an unprotected sand bed around a structure and how effective any scour protection might be. Examples of laboratory investigations carried out by HR Wallingford are listed in Table 1. Elsewhere model tests have been performed on, for example, scaled-down pipelines (e.g. Sumer and Fredsøe, 1990), foundation piles (e.g. Breusers *et al.*, 1977) or other more complex structures (e.g. Sweeney *et al.*, 1988).

Laboratory experiments on scour processes usually make use of a scaled down model of the structure (often simplified to represent the main characteristics) placed on a bed of mobile material in a facility in which waves and/or currents may be generated under controlled conditions. As it is not generally possible to achieve complete dynamic similarity between the model and prototype the results obtained from small-scale model tests should be treated as a guide to prototype behaviour. Despite this limitation, model tests are 'particularly effective for studying scour patterns peculiar to non-standard structural geometries and for optimisation of alternative structural forms to minimise scour at the design stage' (ICE, 1985, p. 76). At the present time it is often considered economically feasible to undertake scaled hydraulic model tests with mobile beds to determine the 3-dimensional scour behaviour for several 'what if' scenarios. Two kinds of scour investigations are performed in the laboratory: (1) process studies where the detailed physical processes and associated scour development around a generic installation are examined and (2) 3-dimensional site specific studies where a geometrically scaled model of an installation is tested to obtain the total scour behaviour.

It is generally assumed (e.g. Carstens, 1966; Breusers and Raudkivi, 1991) that the depth of the scour pit in both the model and the prototype scales with the characteristic length scale of the structure, e.g. the diameter D of a single pile or pipeline.

In many cases sand bed scour occurs under the combined action of waves and currents with constantly varying relative and absolute magnitudes and directions. Despite this the majority of published laboratory investigations are concerned with examining the scour development due to a number of constant wave, current or wave–current flow combinations.

The 2-dimensional modelling generally takes place in laboratory flume channels (typical dimensions: length 5 m to 50 m, width 0·3 m to 3 m, water depths 0·2 m to 2 m) equipped with a pumping system to generate steady unidirectional currents and a wave generating board. In these facilities the combined wave and current flows are limited to flowing in the same direction or in opposition and in all but the largest of facilities the free surface effects often limit the tests to quite a small scale. One solution is to perform the experiments in a duct (i.e. with a rigid lid; cross-section typically 0·5 m by 0·5 m) with waves generated by a servo-controlled piston and pumped steady currents. Such flow ducts are ideal for investigating the bottom boundary layer behaviour and transport of sand under waves and currents but care must be taken to ensure that the results are not artificially influenced by blockage effects due to the presence of the installation, or at least that these effects are understood.

The 3-dimensional modelling studies use square or rectangular basins (typical dimensions: length of sides 20 m to 30 m, water depth 0·2 m to 1 m) with a bank of wave generating paddles situated along one of the basin walls and current generating facilities pumping water across the basin. In these facilities it will generally be possible to vary the angle at which the waves and currents are crossing and the orientation of the structure.

Where scour modelling is performed with long crested waves of regular or random period care needs to be exercised when interpreting the laboratory results as some of the features produced in the model might in reality be smeared by the directional spread of waves in the sea.

If it is only necessary to determine the maximum depth and extent of scour around a structure then model tests can be run with flow conditions for which the ambient bed sediment is just immobile. For a broader coverage of likely scenarios and where information on the live-bed scour development is required, it is necessary to run tests with a net transport of sediment out of the test area. Under these conditions it will be necessary to recirculate the transported sediment or feed new sediment in at the upstream end to replace lost sediment. If the sediment bed is not maintained in the test section by ensuring a continuity in the sediment supply then the results obtained on the scour behaviour may be unrealistic. The effort required to design, install and operate correctly a sediment recirculation loop in the flume or basin should not be underestimated.

Recent advances in specialist techniques (hot-film probe) for measuring the bottom shear stress on fixed beds (Sumer *et al.*, 1992b) have allowed a direct assessment of the scour potential around structures due to waves and currents to be made without the need for mobile bed tests (e.g. Hebsgaard *et al.*, 1994). However, this technique is extremely difficult to use and requires significant experience to obtain good results.

3.2. SCALING CONSIDERATIONS

The scaling of waves and current for laboratory studies is well understood (Froude and Reynolds number scaling) although in practice the many requirements of an experiment preclude an exact scaling exercise. Scale effects may be reduced to acceptable levels in many cases by running tests in suitably large facilities. The (dimensionless) Froude number is defined as

$$\mathrm{Fr} = \frac{U}{\sqrt{gh}} \tag{12}$$

where U is the flow speed, g is the acceleration due to gravity and h is the water depth. The majority of hydraulic models are scaled according to the Froude model law, i.e. prototype Froude number = model Froude number (Hughes, 1993). When the influence of the fluid viscosity is considered to be of possible importance the (dimensionless) Reynolds number is used

$$\mathrm{Re} = \frac{UD}{\nu} \tag{13}$$

where D is a suitable length scale (e.g. water depth or pile diameter) and ν is the kinematic viscosity of the fluid. Viscous effects will be negligible for values of Re based on water depth in excess of 10 000 and the flow can be assumed to be turbulent (Hughes, 1993); below this value the flow will be laminar. As well as achieving a suitable approximatation to the bulk quantities of the flow the vertical structure of the flow may be important. In most cases where the interaction of the sea bed boundary layer with the structure is important the shape of the vertical profile of the current velocity should be simulated correctly.

The relative importance of the following aspects in the model and prototype should be considered when deriving the scale relationship(s) for physical model tests:

- 'traditional' scour–friction factor for bed shear stress, horizontal pressure gradient (ratio of pressure gradient force acting on grain to shear force), threshold of motion, bedload and suspended load, angle of repose for sediment (prototype and model)
- pressure gradient effects both on and in the bed–time-dependent permeable flow (dilation/consolidation), bed density, liquefaction
- rotational slip failure of bed material.

It is important to be aware of, or directly test for, the influence of the model scale on the model results. Tests might need to be run at more than one scale to assess the sensitivity of the results to the various scaling assumptions. The scaling of sea bed (sandy) sediments is for practical reasons (e.g. to avoid the influence of cohesiveness in very fine model sand), not performed geometrically but according to the dimensionless Shields parameter

$$\theta_{cr} = \frac{\tau_{cr}}{(\rho_s - \rho)gd_{50}} \tag{14}$$

where ρ_s and ρ are the mineral density of the bed material and the density of seawater, g is the acceleration due to gravity and

d_{50} the median grain size of the bed material. This means that both the diameter and density of the model bed material are variables that can be selected and as a result both sand and lightweight granular materials can be considered when selecting the model scale.

Bettess (1990) has surveyed (27 questionnaires) usage of the lightweight sediments listed in Table 3 and provides valuable information on the advantages and disadvantages (including difficulty of use) for the various materials. Coal and polystyrene have been used most often (5 replies each) with Bakelite and PVC next (3 replies each), although it appears that Bakelite is no longer manufactured. Bettess comments that whilst lightweight sediments are useful for bedload and initiation of motion studies they are inapplicable where the sediment extends above the water surface, or where the waves are breaking and inertial (pressure gradient) effects are important. Lightweight sediments are often used in wave tank tests because of the Froude scale limitations on the (free-surface) wave conditions. The scaling of sand beaches is commonly based upon a parameter including the settling velocity of the grains and the wave characteristics and the Froude number for the waves (Fowler, 1993; Kraus and MacDougal, 1996). The erosion of accretion of sand beaches without a sea wall can be predicted in terms of the discriminant function H_s/w_sT; for values greater than about 3 the beach tends to be erosional and for values less than 3 the beach tends to accrete. The local significant wave height and period and beach

Table 3. Specific gravity of lightweight materials for use in mobile-bed physical models (after Bettess, 1990)

Specific gravity	Material
1·0 to 1·1	Polystyrene, sawdust with asphalt
1·1 to 1·2	Nylon, Perspex, PVC, wood
1·2 to 1·3	ABS, Bakelite, PVC
1·3 to 1·4	Bakelite, coal, walnut shells
1·4 to 1·5	Bakelite, coal, sand of Loire
1·5 to 1·6	Coal, sand of Loire
1·6 to 1·7	Lightweight aggregate

Note: ABS = Acrylonitrile Butadienne Styrene; Lightweight aggregate = sintered fly-ash, trade name Lytag.

sediment fall velocity are used in the calculation. The physical process behind this rule-of-thumb is that under strong wave action (e.g. in storms) the sediment is suspended in the water column and transported (offshore) by the mean storm-induced current (Hicks and Green, 1997).

There is currently much debate as to the relative merits of using either coal or fine sand for the bed material in coastal models (e.g. Powell, 1987; Loveless and Grant, 1995). Loveless and Grant discuss two approaches denoted (A) and (B) to modelling the sediment transport on shingle beaches. Approach (A) seeks to ensure correct scaling of both the threshold (orbital velocity or shear stress) for sediment motion and the rate of percolation of water into the beach. Approach (B) seeks to ensure the correct scaling of the threshold and the ratio of the percolation forces on the sediment to its submerged weight. Test results using theory (A) (coal) or (B) (sand) produced different results for the same wave conditions, but as the initial beach slopes are slightly different for each sediment and the water depths at the toe of the beach are different the results are difficult to compare. Loveless and Grant suggest a suitable result might be obtained by judiciously varying the angle of repose (angularity of grains) and porosity (percolation rate) of the model sediments but this could be difficult to achieve in practice.

One important aspect when choosing the model sediment concerns the formation of ripples. In laboratory tests with quartz sand the ripples will often be large in comparison with the scaled model installation and also produce an unrealistic bed friction regime. The frictional resistance of lightweight sediments in a simulated tidal flow has been investigated experimentally by Bayazit (1967), for sand, coal, Perspex and wood. If it can be assumed that bed friction scale effects are not important overall, the effect of ripples on the actual scour development is often found to be negligible, e.g. in model tests of scour by waves at a vertical pile (Sumer *et al.*, 1992b). This is because the enhanced bed shear stress and turbulence levels adjacent to the model structure cause the ripples to be locally washed out, whilst elsewhere in the model the ripples can be treated as a periodic 'noise' around the mean bed elevation. In cases where the flow speed is very large it may be possible to use a coarse sand for the bed material in the model; this is advantageous as sand larger than about 0·8 mm does not ripple easily. For example, HR Wallingford used a uniformly graded Leighton Buzzard sand with a mean sediment diameter of 0·8 mm for studying local scour around a barrage in the Taff Estuary (South Wales). The geometric model scale was 1:50.

The influence of bed material size on laboratory results must be considered because when the d_{50} of the bed material is large compared with the diameter of the pile D this will reduce the scour depth. The available data indicate that there is no influence of the particle size on the scour depth providing d_{50} is smaller than $D/25$ but preferably $D/50$ (Breusers and Raudkivi, 1991). The influence of sediment grading should also be considered as there is some evidence that this may exert a control on the scour depth development (Breusers and Raudkivi, 1991).

Sediment particle shape might also be considered. Clark *et al.* (1982) obtained some uncharacteristically small scour depths in laboratory tests with lightweight sediment and indicated that this was due to a small percentage of large grains in the bed material that had not been removed when preparing the bed and, possibly, due to the shape of these grains which were flatter than natural sand.

Finally, some consideration is required regarding the mode of sediment transport. In the sea the suspended sediment transport rate is often the dominant mode of transport (Soulsby, 1997) and hence the relative contributions to the total transport rate from bedload and suspended load in laboratory tests should be also considered when interpreting the results. Irie and Nadaoka (1984) – in Oumeraci (1994b) – investigated the scour pattern in front of a sea wall and found a dependency on the ratio U_w/w_s which contains the maximum bottom orbital velocity due to the wave action U_w and the settling velocity of the bed grains w_s. For approximate values of $U_w/w_s \geq 10$ the scour pattern is as for a relatively fine sediment (suspension dominated transported) whereas for $U_w/w_s < 10$ the sediment is relatively coarse (bedload dominated transport) (see Section 7.6.1). The implications of the mode of sediment transport on the results of model tests are discussed in depth by Hughes (1993).

3.3. MODEL DESIGN

In summary the following points need to be considered when setting up the model and whilst interpreting the results:

- is it important to include sediment transport processes away from the area influenced by the structure?
- is information on the rate of scour development required in addition to the location or depth and extent?
- are geotechnical (bed seepage, pore water pressure, soil failure) processes likely to be significant in the prototype situation? The scaling laws for these processes are different from those relating to sediment transport and scour
- how important are time-varying factors (wave climate, tidal current speed, direction of flow/waves, water level)?
- are the geotechnical aspects of the foundation behaviour adequately represented?, e.g. with respect to settlement of foundations
- is the ratio of the design flow condition to the threshold of scour scaled correctly?
- is the surface roughness of the model likely to be a significant factor in determining the vortex shedding or drag on the face of the structure?
- breaking waves – are breaker effects adequately represented?
- relative importance of 2-dimensional vs 3-dimensional effects
- relative importance of long crested waves vs short crested seas
- laminar vs turbulent boundary layer effects
- relative importance of wave and current
- relative importance of bedload and suspended load transport.

Other factors:

- bedforms – rippled sediment can produce unrealistically scaled bed friction and 'noise' on the scour development
- artificially high blockage in the laboratory model can be avoided with

$$\frac{\text{Area (model)}}{\text{Area (flume)}} \leq \frac{1}{6}$$

where Area = cross-sectional area projected to flow
- water depth effects can be avoided if $h \geq 4D$ where h is the water depth and D the characteristic dimension of the structure, e.g. pile diameter

- widely graded sediments or sediment particle shape – erroneous results might be obtained if 'platy' particles armour the bed surface or if the maximum grain size exceeds $D/25$
- cohesive sediments (or cohesive fraction in sandy sediment) – the presence of greater than 10% by mass of cohesive sediment alters the threshold of scour and scour pit shape. A rule of thumb is that bed material with a median grain diameter less than 0·100 mm is likely to exhibit some cohesive behaviour.

3.4. INTERPRETING THE RESULTS

3.4.1. *Semi-empirical models*

The laboratory data are often used to produce semi-empirical predictive equations for scour development. Suitable dimensionless equations can be derived to predict scour development within the range of validity of the scale data while paying due attention to the quality of the original data, goodness of fit, etc. It is best to use the principles of physics to derive dimensionally homogeneous equations. However, the introduction of 'plausible' dimensional quantities (e.g. gravity, viscosity) without a physical basis should be avoided, and it can lead to erroneous deductions. The extrapolation of the results by a scale-series outside the intended range should be undertaken with extreme caution and only after considering whether the processes acting in the physical model tests are still acting in the prototype situation.

3.4.2. *Numerical models*

The numerical modelling of scour development is reviewed separately in Chapter 4 but as with the physical modelling there are options of 2-dimensional or 3-dimensional flow modules, and the most appropriate needs to be chosen. The sediment transport modules often adopt a relationship for predicting the transport rate which is driven by the shear stress τ_0 but one important consideration when modelling scour downstream of an installation is how to deal with flow which has separated around the installation. Firstly, the shear stress vector fluctuates in time producing enhanced turbulence levels, and an associated increase in the sediment transporting capacity of the flow, and secondly at the point of reattachment, where the separated flow impinges back onto the bed, the time-mean value of τ_0 is zero, i.e. zero net transport.

The numerical model must represent satisfactorily some or all of the most important flow–sediment linkages. However, even a basic model may be used to analyse several 'what if' scenarios in conjunction with a physical model study ('hybrid' approach). The numerical model can provide information on the general behaviour of the flows and sediment and the boundary conditions for the physical model, and the physical model is then used to assess the detailed scour behaviour.

3.4.3. *Field measurements*

Field data provide the ultimate validation for the results of scale model tests (and numerical modelling studies). However, owing to the difficulty of obtaining data in severe conditions, most field data on scour are collected in the summer months, when the wave–current climate is less severe and the sediment mobility and scour are reduced.

From a practical point of view, given the reliability of the flow and water-level sensors currently available and recent advances in sonar techniques for measuring the shape of the bed around structures (e.g. Ingram, 1993), as well as some more simple devices (e.g. Waters, 1994) for monitoring scour, the continuous monitoring of the flow field, sediment load and scour morphology at the periphery of permanently installed structures should not present insurmountable practical problems.

Whilst case studies are extremely valuable to the engineer who is trying to assess the capabilities of the various available scour prediction methods or accuracy of model results, most observations of severe scour problems are often commercially confidential, or at least sensitive, and will not generally be reported in the open literature.

Computational modelling of local scour

4. *Computational modelling of local scour*

4.1. INTRODUCTION

Over the past ten years computational (numerical) models have begun to be used to make scour predictions. However, the number of studies is not large, and those reported deal with simple geometries (cylindrical pipelines) and non-cohesive beds (evenly-graded sand) in simple (steady current) flows. Even in idealised conditions like this, the problem is by no means a trivial one. It involves a turbulent shear flow and its interaction with an erodible moving bed, usually the flow field is complex because it is disturbed by the structure and it also carries suspended sediment. The results obtained so far are encouraging in the sense that the scour hole depths and overall shape that have been predicted are in reasonable agreement with laboratory experiments. Some previous work is described below in which Sections 4.3 and 4.4 are based upon an earlier review by Brørs (1993).

4.2. GENERAL APPROACH TO NUMERICAL SCOUR CALCULATIONS

The basic procedure for making scour calculations is illustrated in Figure 8 and the bed change is determined according to the sediment budget equation or continuity equation (see Equation (1)). This equation states that the change in bed level is related to the divergence in the sediment flux field.

Information on various approaches to making morphodynamic predictions with numerical models has been given by

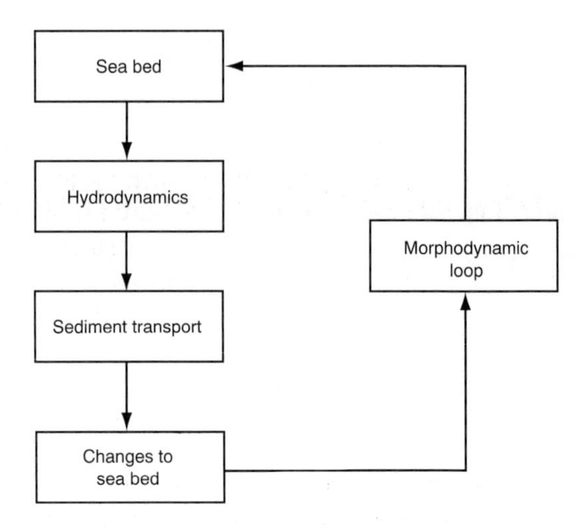

Figure 8. Flow diagram for making scour predictions with computational models

Chesher *et al.* (1993) and de Vriend *et al.* (1993). These authors discuss ways of treating the bed updating required in morphodynamic models and how to cope with the separate time-steps for the flow modelling and bed updating (morphodynamic loop).

4.3. EXAMPLES OF NUMERICAL SCOUR CALCULATIONS

A number of complex 2-dimensional scour numerical models have been reported based upon a description of the turbulent flow around a pipeline. Whilst the flow modelling is quite well refined (e.g. Van Beek and Wind, 1990) the link with the sediment transport and scour is less good especially in the region where the separated boundary layer flow reattaches to the bed downstream from the pipeline.

Breusers (1975) made an early numerical study of the velocity field in the scour hole but did not compute the sediment transport rate and scouring. Almost ten years later Leeuwenstein and Wind (1984) published predictions of scour under a pipeline using the Navier–Stokes equations coupled with a k–ε turbulence model. The production of turbulent kinetic energy k and its dissipation rate ε were solved from two energy transport equations. The sediment transport calculations, performed using a shear stress driven bedload formulation, reproduced the general scouring behaviour but there were some apparently artificial results such as ripples forming and being transported under the pipeline. Van Beek and Wind (1990) applied the same flow model with both bedload and suspended load sediment transport formulations. Comparisons made with the data of Kjeldsen *et al.* (1974) were promising although the erosion time was much shorter in the numerical model.

Recent experiences with $k-\varepsilon$ modelling of flow and pipeline scour at HR (Brørs, 1997) have indicated that the numerical modelling of 2-dimensional pipeline scour can reproduce reasonably well the bed evolution under the pipe, and predict maximum scour depths in reasonable agreement with laboratory scour data, within 20%. However, the computing time required is still large and prohibitively so for 3-dimensional problems (see Table 4).

Other approaches include adopting a hybrid modelling approach based upon field and laboratory results. For example, Staub and Bijker (1990) presented a dynamical numerical model for predicting the sandwave characteristics and pipeline self-burial due to tunnel erosion, leeside erosion and backfilling. This type of model is more practical to operate than a full flow modelling approach as the scour and self-burial are parameterised from data and the numerical model is used to predict the rate of change.

4.3.1. Scour downstream of a rigid bed

The work of Puls (1981) was aimed primarily at the numerical simulation of bedforms, but a case of local scour downstream of a rigid (protected) section of bed was also predicted. The flow configuration was two-dimensional (vertical profile) steady flow and the Reynolds-averaged Navier–Stokes equations with (k–ε) turbulence closure were solved using finite differences. Bedload and near-the-bed suspension transport was modelled in terms of a friction velocity u_* dependent local erosion rate, and a corresponding local deposition rate expressed in terms of a friction velocity dependent local probability for deposition. A random walk particle method was devised to account for sediment leaving the bed in the recirculation zones downstream of ripples. The predicted scour was too slow to start with and became too fast later, compared with laboratory measurements. The predicted scour hole depth compared well with the laboratory experiment, but the predicted scour hole length to depth ratio W/D of about 30 was too large compared to the value of about 10 found in experiments.

Hoffmans and Booij (1993) studied the scour hole development downstream of a bed protection layer. Their model used the hydrostatic pressure assumption. The momentum equation and the convection–diffusion equation for suspended sediment concentration were solved using an algebraic expression for the eddy viscosity (zero-equation turbulence model). A stochastic approach was applied for specifying the reference bed concentration as well as for the calculation of bedload transport. The predicted scour was reported to be acceptable compared to both a scale model and a prototype case, although the predicted scour hole development appeared to be too fast.

Recently (Ushijima, 1995) 3-dimensional numerical scour predictions using a turbulence model have been obtained for the case of local scour due to the flow from a cooling water outfall.

4.3.2. Two-dimensional (pipeline) scour

Leeuwenstein and Wind (1984) presented a model based upon $k-\varepsilon$ turbulence and a transport equation for sediment concentration, although the latter was not used in the calculations which only included bedload transport modelled by means of conventional bedload formulae. In the scour development predictions, a shallow initial scour hole was used in the initial flow calculation. Leeuwenstein and Wind reported that in the calculation, a ripple forms at the bed below the pipeline and moves downstream. This feature was not observed in laboratory experiments, and is believed to be caused by the bedload formula being a too rigid schematisation for the case of sediment transport encountered in local scour. The authors suggested four possible ways to circumvent the problem: to include suspended load transport, to introduce the stochastic nature (variance) of the bed shear in some way, to account for variable bedform-induced bed roughness in the near-bottom boundary conditions, or perhaps to use a revised bedload formula.

Sumer *et al.* (1988a) applied a discrete vortex (cloud-in-cell) model and demonstrated that the model was able to represent the vortex shedding process and general flow pattern in the lee wake of a cylinder above the bed, with and without a scour hole present. No actual computations of scour were performed, but the authors made the important conclusion that the time-averaged bed shear stress is not a suitable parameter to work with in mathematical modelling of the lee-wake erosion behind a pipeline.

Van Beek and Wind (1990) predicted the scour development scour holes beneath a pipeline, with and without an attached spoiler, using a flow model with k–ε turbulence closure and a transport equation for suspended sediment. Although the bed changes are apparently calculated only from vertical sediment fluxes at the bed (suspended load transport), good agreement is demonstrated with measured scour holes. The rate of erosion was however reported to be three times as fast in the calculations as in the physical model. The authors concluded that in further work the effects of sediment-induced stratification on the flow should be taken into account, and that some improvements to the numerical scheme could be made.

Brørs (1997) used a Taylor–Galerkin finite element flow and sediment transport model (bedload and suspended load) with k–ε turbulence. Density effects were included in the (vertical) momentum equation and in the turbulence equations. Periodic vortex shedding from a cylinder placed slightly (gap $0{\cdot}5D$) above a rigid bed was predicted in good agreement with laboratory experiments of Sumer *et al.* (1988a). However, a fine mesh was needed in the vicinity of the cylinder in order to obtain periodic shedding – 50 nodes around the perimeter and 5–6 nodes through the boundary layer developing on the cylinder surface proved to be sufficient, giving a grid of approximately 5000 nodes in all. The model domain covered approximately $22D$ in the flow parallel direction and $7D$ in the vertical. The scour calculation was performed using coarser ($\sim$3000 nodes) grids. Small-scale bed features (as observed by Leeuwenstein and Wind, 1984) tended to develop in the calculations, limiting the morphological time steps to 1–2 seconds for a 0·1 m diameter pipe in a steady flow with mean velocity $0{\cdot}35\,\mathrm{m\,s^{-1}}$. Figure 9 shows the computed maximum scour depth and deposition downstream as well as scour hole profiles obtained in the laboratory by Mao (1986) and Kjeldsen *et al.* (1974). The effect of vortex shedding was included (from $t \approx 13$ minutes) by increasing the predicted bed stress in the lee wake region between $x = 0D$ and $x = 7D$ with a peak enhancement of 24% at $x = 3{\cdot}5D$. The predicted scour development and bed profiles compare well with the laboratory measurements for Mao ($\theta = 0{\cdot}048$; clear water scour). The results from this study suggest that the (at least parameterised) influence of vortex shedding needs to be included to predict correctly the scour depth, but the width of the scour pit is overestimated both with and without the inclusion of this effect. Further details of the numerical model are given in Appendix 1 and in Brørs (1997).

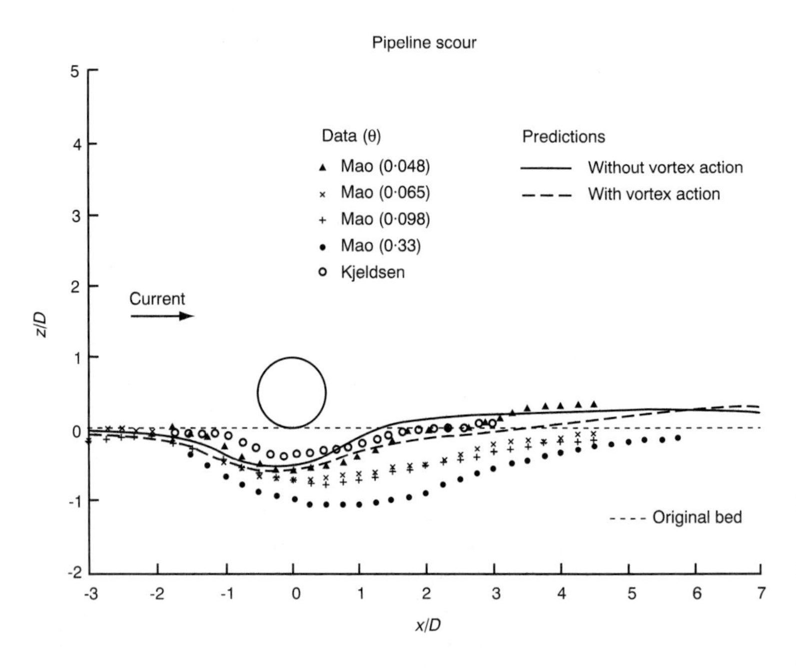

Figure 9. Predicted time development and bed profiles for scour at a pipeline compared with measurements by Mao (1986) and Kjeldsen et al. (1974) (figure provided by Brørs, B. of SINTEF, Civil and Environmental Engineering, Trondheim)

4.3.3. *Three-dimensional scour at a vertical cylinder*

Olsen (1991) and Olsen and Melaaen (1993) calculated scour at a large diameter vertical cylinder in steady flow using a finite-volume model with sediment transport equation and k–ε turbulence. The simulated situation was a laboratory experiment (Torsethaugen, 1975) of scour at a vertical cylinder with a diameter $D = 0{\cdot}75$ m, in a flow with depth 0·33 m and upstream velocity $U = 0{\cdot}067\ \mathrm{m\,s^{-1}}$, and with the bed consisting of $d = 3$ mm plastic particles having relative density $s = 1{\cdot}04$. The calculations were carried out for a chosen time (i.e. until the numerically predicted 'maximum depth in the scour hole was equal to the observed scour hole depth in the physical model study'), and no equilibrium scour depth or time history of scour development is reported. As can be seen in Figure 10, the

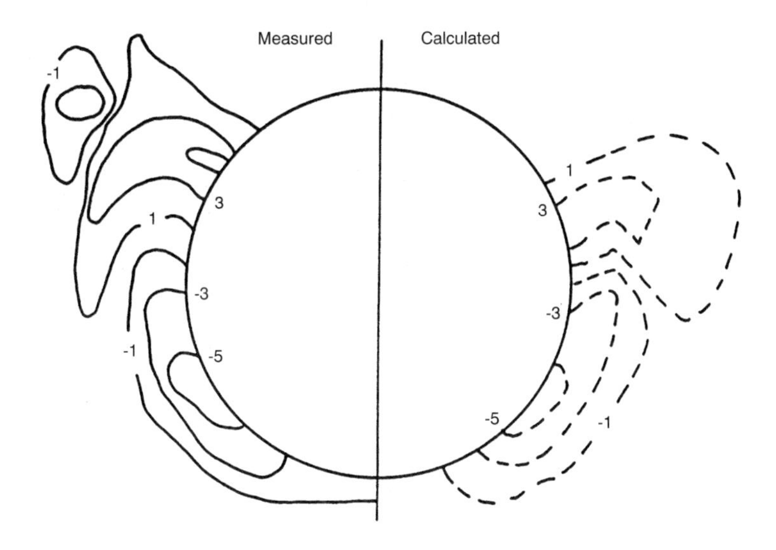

Experimental conditions for measured result:

cylinder diameter = 0·75 m
relative sediment density = 1·04
sediment particle diameter = 3 mm
water depth = 0·33 m
ambient flow speed = 0·067 m s^{-1} (clear-water scour)

Figure 10. Predicted scour and accretion at a vertical cylinder compared with measurements by Torsethaugen (1975), scour contours at 2 cm intervals (reproduced from Olsen and Melaaen, 1993, by permission of the ASCE)

calculated scour hole geometry agrees well with the experimental data. Recently a numerical prediction of the scour development around a slender vertical cylinder in a steady flow has been completed (Olsen, N. R. B., personal communication, 1996).

A number of Japanese workers have published the results of numerical and experimental studies of the flow and scouring around large circular cylinders in waves and currents (Saito and Shibayama, 1992; Toue *et al.*, 1992; Saito *et al.*, 1990). Although the general patterns and magnitudes of erosion and accretion have been predicted the detailed agreement is generally not perfect.

4.4. CONCLUSIONS RELATING TO TURBULENCE MODELLING OF SCOUR

One important factor that decides whether or not numerical scour calculations are feasible is the computing time. Table 4 is based on experience at HR Wallingford during 1993–94 (after Brørs, 1997) with the numerical prediction of 2-dimensional flow and scour at pipelines. The computing times are estimated for a Sun 10 workstation, and the estimates are for calculations until equilibrium scour is achieved. N_{bed} are the number of bed updates required and N_{flow} are the number of intermediate flow updates before each bed update is made. The fine grid captures vortex shedding, and 1250 time steps are necessary per vortex

Table 4. Assumed computing time for calculation of equilibrium scour profile under a pipe in steady flow (after Brørs, 1997)

	Medium grid (3000 nodes)	Medium grid (3000 nodes)	Fine grid (5000 nodes)
Time-steps ($N_{bed} \times N_{flow}$)	6000×200^1	20×200^2	40×1250^2
Flow time-step	0·25 s	0·25 s	0·4 s
Comp. time, 2-dimensional	83 h	0·3 h	5·5 h
Comp. time, 3-dimensional	2+ months	5+ h	100+ h

[1]Bed update scheme with 'ripple' formation, present model.
[2]Assumed 'perfect' bed update scheme.

shedding period. The medium grid produces a steady flow field and 200 time steps are sufficient for each grid update. The estimates for 3-dimensional are simply made by multiplying with a number (20 +) of 'layers'.

So far, only the simplest of flow conditions and bed materials have been attempted. Tidal flow situations (Keulegan–Carpenter number > 300) can probably be simulated by consecutive steady flow runs with changing flow direction, and would be feasible on a workstation with a 'perfect' bed update scheme.

Olsen and Melaaen (1993) reported that they used ten iterations on a Cray supercomputer for their three-dimensional cylinder case, equivalent to between 100 and 150 minutes' computer run time (Olsen, N. R. B., personal communication, 1996). Recent numerical predictions of the scour development at a slender pile took over 1 month to run on an IBM 370 machine (Olsen, N. R. B., personal communication, 1996).

Hence, at present it appears that even to attempt the practical numerical modelling of scour for the uniform 2-dimensional case, let alone more complicated flow and bed conditions, will require improved modelling techniques and increased access to faster machines. There is some evidence of an improvement in the latter as the pipe scour model referred to above developed in 1993–94 is now running on a PC (Brørs, B., personal communication, 1997).

4.5. MORPHODYNAMIC CHANGE IN DEPTH-AVERAGED MODELS

Recent morphodynamic investigations, using a depth-averaged flow coastal area model, of the beach response to waves to the lee of a detached coastal breakwater (e.g. Price *et al.*, 1995) show that locally the bed at the two ends of the breakwater is lowered in a fashion that looks remarkably like the development of local scour (Figure 11).

The result illustrated in Figure 11 was based on the work of Price *et al.* in which the depth-averaged *2DH* flow equations were linked to a morphodynamic model, via an energetics based sediment transport formula, to predict the bed development. The input conditions were as follows:

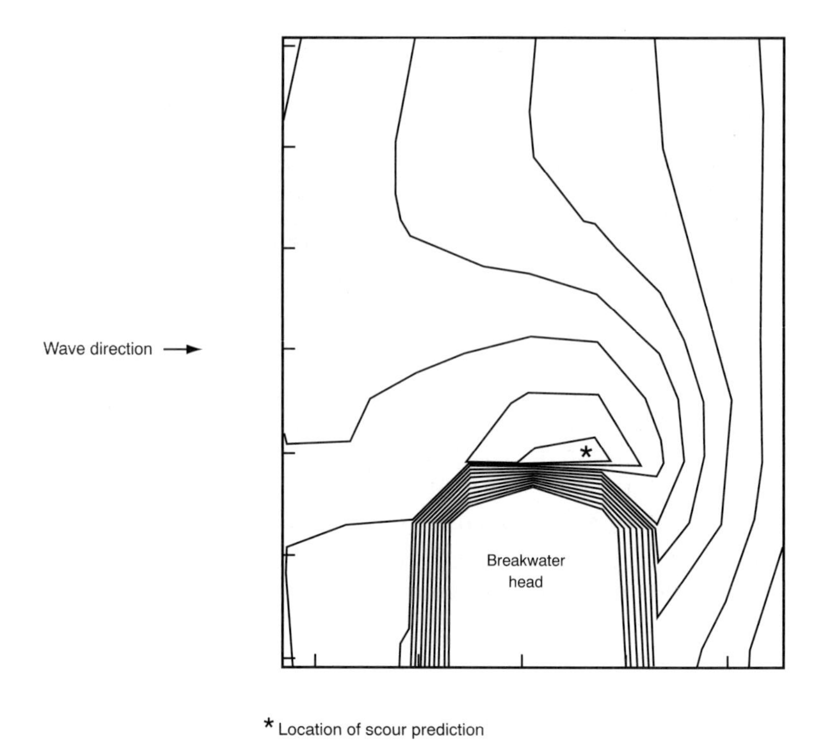

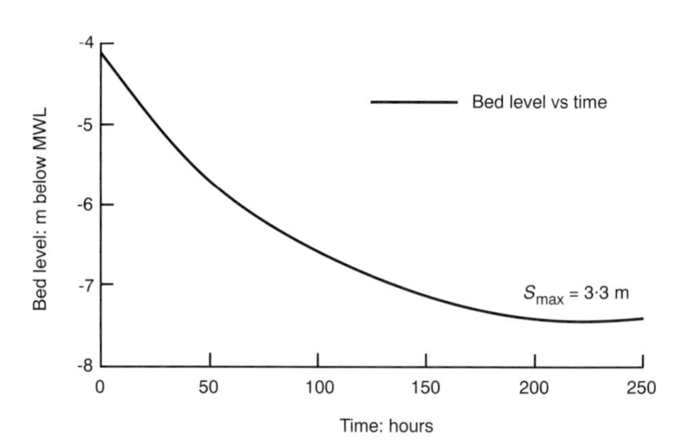

Figure 11. Scour development around a detached breakwater predicted by a depth-averaged flow morphodynamic model (after Price et al., 1995)

Offshore water depth	7 m
Root-mean-square wave height	2 m
Peak period of wave spectrum	8 s
Angle of attack of waves	0° (wave crests parallel to bed contours)
Beach slope	1:50
Bed material d_{50}	250 μm
Width of breakwater	15 m
Length of breakwater	300 m

The model was run with the breakwater placed in an initially plane beach bathymetry to simulate 250 hours of wave action with a mobile bed offshore ($\theta > \theta_{cr}$). The bed contours around the breakwater were deformed and the scour hole at the tip of the breakwater developed to a maximum equilibrium depth of around 3·3 m, or 20% of the breakwater width. Comparing this result with the model tests of Gökçe *et al.* (1994), for the present configuration KC number is less than 1 and for this situation the results of Gökçe *et al.* would predict that the scour depth is negligible due to wave action alone. However, the wave action in the breakwater case modelled by Price *et al.* produces a circulation to the lee of the breakwater and a return flow of water around the tip. Here, and as found by Gökçe *et al.*, one would expect the wave induced scour to be enhanced by the presence of even quite a small current. The presence of the current probably accounts for the 3 m scour depth observed in the computer model.

However, the scour predictions obtained from coastal area models (de Vriend *et al.*, 1993), solving the depth-averaged (2*DH*) flow equations and using an energetics based sediment transport model, and other 2*DH* simulations have not yet been validated with laboratory or field data. Therefore the accuracy of this kind of approach to predicting scour at structures is not at present clear, although it probably gives a good indication of the scour behaviour.

4.6. CONCLUSIONS

A number of studies over the last 10 to 15 years have successfully predicted the flow pattern and turbulence characteristics for the idealised 2-dimensional pipeline case (vertical slice through pipeline) with and without the presence of a scour hole. The more complex task of using the turbulence models in a fully morphodynamic fashion to predict the scour development under the pipeline has received less attention, but this can be done reasonably accurately for the case of a steady, uniform flow. The modelling of scour development with 3-dimensional codes (e.g. vertical piles) has started to receive attention more recently and it appears that reasonable results can be achieved.

Owing to the prohibitively long computing times required for relatively simple turbulence modelling of scour for both 2-dimensional and 3-dimensional cases, combined with the lack of a robust approach (the models require continuous intervention), it is probable that this method will remain as a research tool for the foreseeable future. However, it should not be ruled out that numerical modelling could be routinely used for engineering studies of local scour development within the next ten to fifteen years, based on current trends for the expected increases in computing power and the development of more efficient numerical schemes.

The coastal area models are feasible to use in engineering studies although the accuracy of the scour evolution predicted for coastal structures by these models, which were primarily developed to predict the morphodynamic response of the coastline, has yet to be verified.

The wave–current climate

5. The wave–current climate

5.1. INTRODUCTION

Where structures are founded in potentially mobile sediments it is important to quantify the degree of mobility and hence scouring of those sediments. Because a wide variety of installation types are placed in the marine environment a generic methodology for assessing the potential for scour is required. The aim of this chapter is to provide a framework with which to assess the mobility of the sediments and the likelihood for scour. The methods described can be used to predict whether the sediment is mobile around a specific structure under design conditions, or alternatively to estimate the probability of sediment becoming mobilised under a specified wave–current climate.

The methodology adopted here builds on the general approach to scour discussed in Chapter 2 which can be summarised as follows. The effect that the structure has on the ambient, undisturbed flow field determines whether sediment becomes mobile at the toe of the structure. The degree of disturbance imparted to the flow depends upon the shape, size and orientation of the structure, and the flow disturbance in turn controls the magnitude of the shear stress amplification and its distribution over the sea bed. The shear stress amplification is the principal factor determining the onset and the resulting pattern of scour as the locally larger values of shear stress enhance the capacity of the flow to transport sediment. This in turn leads to a local divergence in the sediment transport rate producing a scour depression at the toe of the structure.

The methodology assumes:

(*a*) that the mobility (or not) of sediment on an area of the bed uninfluenced by the structure can be determined using a conventional sediment transport approach, i.e. the bed shear stress τ_0 (ambient) controls the rate at which sediment is transported across the sea bed. This underpins the whole approach to calculating the potential for scour.

(*b*) the mobility of sediment adjacent to the base of the structure (even if sediment is immobile elsewhere) can be predicted as an extension to (*a*), through use of the coefficient M which describes the shear stress amplification at a given position adjacent to the structure. M is defined as the ratio τ_0 (local to the structure)/τ_0 (ambient) and typical values lie in the range 4 to 9 depending on the shape and size of the structure, and the orientation of the structure to the flow direction (Section 2.5.1). If, therefore, the sediments are immobile at a position close to but uninfluenced by the structure, then local scouring of the same sediments at the structure may occur (clear-water scour). If the ambient sediment is mobile in the area, then local scour will certainly occur (live-bed scour).

The methods presented indicate the likelihood for scour occurring but they do not directly indicate the local depth of scour that will occur because this is controlled by the combined dependency of the equilibrium scour depth and the rate of scouring on the bed shear stress. A worked example involving the calculation of the likelihood for scour is included in Section 5.7.

Although references to data sources relate to the UK continental shelf, the techniques described can be used at any location.

5.2. SEDIMENT MOBILITY ON THE CONTINENTAL SHELF

Although the mobility of sediment varies both spatially and temporally the sea bed sediments on the continental shelf are in long-term (at least decadal) equilibrium with the prevailing hydraulic climate. Gravel (or rock) substrates correspond to those areas experiencing the strongest currents, whilst muddy sediments correspond to weak currents and the areas of sand, which occupy over half the area of the north-west European continental shelf, correspond to intermediate spring tide surface current strengths of typically 0·3 to 0·8 m s^{-1}. The occurrence of bedforms in cohesionless sandy sediments is also closely linked to the strength of the maximum spring tidal current. Consequently, the simultaneous occurrence, for example in the southern North Sea, of mean spring near surface currents greater than 0·5 m s^{-1} with an abundance of mobile sandy sediments leads to the generation of sandwaves and other bedforms (Stride, 1984; Dyer, 1986). If unchecked, scour around structures is very likely to occur in such environments but can also occur in less energetic areas due to the localised shear stress amplification.

In a weak flow no sediment movement will take place but as the flow strength is increased a few grains will gradually begin to move and above a certain flow strength the sediment will be generally mobile. Sediment particles located on the sea bed are moved by the tangential fluid force or bed shear stress, $\vec{\tau_0}$, due to currents, waves, or both, which is usually assumed to take the form:

$$\vec{\tau_0} = \rho C_r \mid \vec{u} \mid \vec{u} \tag{15}$$

where ρ is the fluid density, C_r is a dimensionless resistance coefficient related to the surface texture of the sea bed, and the flow strength, and $\vec{u}$ is the instantaneous fluid velocity vector at some distance above the bed. The vertical variation in the current velocity near the sea bed and the magnitude of C_r are dependent on the texture of the sea bed, commonly identified by z_0 the hydraulic roughness length (Soulsby, 1997).

If the flow is strong enough to move sediment the sea bed will often be covered by ripples with a typical wavelength of 0·2 m and a typical amplitude of 0·1 $\times$ wavelength. The presence of ripples (a likely situation) enhances the bottom friction felt by the flow because of the combined contribution of the drag arising from the individual sand grains (skin friction, related to the median grain size d_{50}) with the (much greater) form drag arising from the variation in the bottom pressure field over the ripple topography, i.e.

$$\tau_0 = \tau_{0s} + \tau_{0f} \tag{16}$$

where τ_{0s} is the skin friction contribution and τ_{0f} is the form drag. The identification of these values is important since only the skin (grain related) friction can move sediments. For future reference the subscripts s and f are dropped, since all shear stresses will be calculated using the assumption that τ_0 is determined by the sediment grains, i.e. using z_0 with d_{50} as the length scale, the grain related bed shear stress.

A flow chart of the procedure for predicting whether sediment is mobile at the particular sea bed location, and how to include the effect of the structure when assessing the likelihood for scour, is presented in Figure 12. The methods used for calculating the bed shear stress and the threshold of motion for cohesionless sediments are described in Appendix 2. A fuller account of these methods and their derivation is contained in Soulsby (1997).

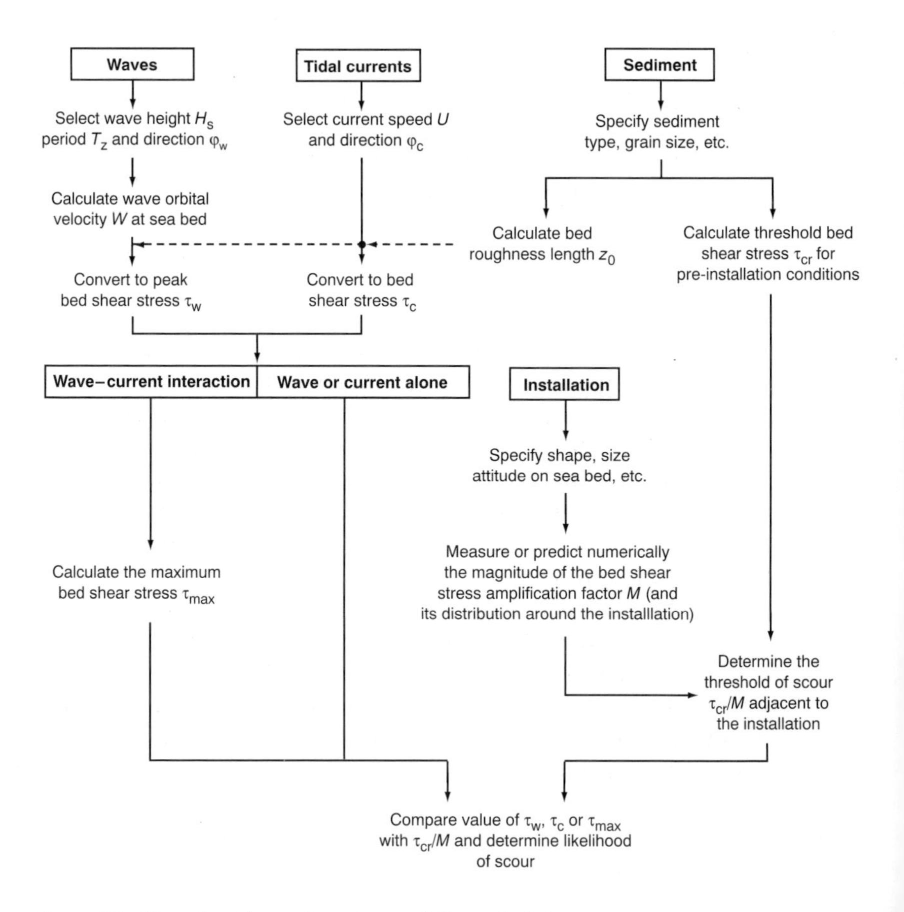

Figure 12. Flow chart for sedimentary mobility calculations

5.3. ENVIRONMENTAL INPUTS

The three environmental inputs required for making mobility calculations are:

- currents
- waves
- sediments.

In the situation where waves and currents are both present the calculations should take account of the wave–current interaction.

5.3.1. *Currents*

Representative values of the surface tidal current velocity can be determined at specific locations from the tidal diamonds present on Admiralty charts. A more extensive coverage, at a coarser resolution, is available in the Admiralty tidal streams atlases or the atlas of Sager and Sammler (1968), and its derivative (BODC, 1991). The magnitude and direction of the depth-averaged flow of an average spring tidal current are plotted in Department of Energy/Health and Safety Executive (1992/93). If there is likely to be a significant non-tidal component of current (e.g. storm surge) the magnitude of the non-tidal component can be assessed from Department of Energy/Health and Safety Executive (1992/93, 50 year return depth-averaged storm surge current speed). Noble Denton (1984) plots the distribution of the extreme total surface current. Where tidal flow information is required close to the coastline, and in the absence of a tidal diamond, the output from a computational model of tidal flow for that location should be used, although in some instances it might be considered preferable to use the data obtained from field measurements. The locations at which current meter records have been obtained and are held by the British Oceanographic Data Centre are indicated in the UK Digital Marine Atlas (BODC, 1991).

The motion of the seas due to tidal currents exerts a frictional force (drag) on the sea bed sediments which can be expressed in terms of the current-only bed shear stress τ_c. The drag exerted on the flow of water by the sea bed produces a vertical variation in the current speed with a velocity profile which is characterised by an increase in speed with distance from the sea bed. The thickness of the bottom boundary layer associated with this velocity variation under oscillatory (tidal or wave) flow can be described in one of two ways:

- the distance from the bed to which the vertical variation in flow speed is described by a logarithmic velocity profile (Soulsby, 1990), or
- the height above the bed where the turbulent kinetic energy in the flow goes to zero, i.e. where the flow no longer 'feels' the bed.

The boundary layer thickness is an important parameter when considering scour processes around marine structures because of the vertical variation in the dynamic pressure acting on the leading edge of that part of the structure within the boundary layer.

Under a rotary tidal flow, as occurs in the open sea, the flow speed is never equal to zero and the boundary layer always has a finite thickness which varies over the UK continental shelf as described in a Department of Energy/Health and Safety Executive (1992/93) publication or by Soulsby (1990). In the deep water of the northern North Sea (north of 54° N), for example, the boundary layer is typically 20 to 40 m thick and the vertical distribution of the current velocity can be described in terms of the complex physics of planetary boundary layer flow (Department of Energy/Health and Safety Executive, 1992/93; Soulsby, 1990). However, for calculations which involve only the current speed a simple power law formula can be used which has been shown to agree well with measurements from many locations (Soulsby, 1990; Whitehouse, 1993)

$$U(z) = \left(\frac{z}{0{\cdot}32h}\right)^{1/7}\bar{U} \qquad 0 \leq z \leq 0{\cdot}5h \tag{17a}$$

$$U(z) = 1{\cdot}07\bar{U} \qquad 0{\cdot}5h \leq z \leq h \tag{17b}$$

where $\bar{U}$ is the depth-averaged current speed, $U(z)$ is the current speed at height z above the bed and h is the water depth.

5.3.2. *Waves*

Design wave characteristics can be obtained from Department of Energy/Health and Safety Executive (1992/93, estimated 50 year return significant wave height) or Noble Denton (1984, 50 year extreme maximum wave height and associated crest to crest period). Maps of the seasonal and annual representative wave parameters exceeded 75%, 50%, 25% and 10% of the time are depicted in the comprehensive wave atlas produced by Draper (1991). Alternatively, and especially for nearshore sites where shallow water effects will become important, the wave climate can be determined directly from the analysis of measurements, or through numerical model results based upon the hindcasting of waves from long records of the wind speed and direction at a nearby location. The locations at which wave recordings have been obtained in the past are indicated in the UK Digital Marine Atlas (BODC, 1991) and a catalogue of near-shore locations around the UK for which synthetic wave data have been generated is available (Harford and Ramsey, 1996).

The bottom orbital velocity generated by the passage of wind waves of height H_s and period T_z over the sea surface generates an oscillatory bed shear stress. The amplitude of the bottom orbital velocity varies with wave height H_s, wave period T_z and water depth h and a method for calculating the bottom orbital velocity for a spectrum of sea surface wave heights and frequencies has been devised (Soulsby, 1987). The value of the maximum bed shear stress τ_w in the wave cycle determines whether the sediment is mobilised by the waves.

The wave boundary layer is generally much thinner than the current boundary layer. Laboratory measurements (e.g. Jonsson and Carlsen, 1976) in an oscillating water tunnel with a rough bed have demonstrated a maximum boundary layer thickness in the wave cycle of between 0·02 and 0·10 m depending upon the maximum bottom orbital velocity U_w, wave period T_z and the bed roughness z_0. Numerical models (e.g. Davies *et al.*, 1988) and analytical models (e.g. Myrhaug and Slaattelid, 1989) predict thicknesses of between 0·10 and 0·15 m. One important consequence of the wave boundary layer being much thinner than the current boundary layer is that the shear stress exerted under a wave with a given value of U_w is much larger than for a current of the same velocity.

5.3.3. Wave–current interaction

In wave–current boundary layer flow as the wave action increases the mean current speed near the bed is progressively reduced, whilst the shear stress is increased. In most cases and especially under wave dominated conditions there is a significant amount of non-linear interaction between the wave and current boundary layers which enhances the maximum bed shear stress τ_{max} in a wave-current flow by as much as 50% (Soulsby *et al.*, 1993). This results in values of τ_{max} which are significantly larger than are found by a simple linear addition of τ_c and τ_w., i.e.

$$|\vec{\tau}_{max}| \geq |\vec{\tau}_c| + |\vec{\tau}_w| \quad (18)$$

where subscripts wc, c and w denote the bed stress due to combined waves and current, waves and currents, respectively. The maximum shear stress τ_{max} can be enhanced by more than 50% and the mean current speed can be reduced by up to 50% with relatively strong wave action. Results indicative of the reduction in mean current speed at a height of 0·25 m above the bed are reported by Slaattelid *et al.* (1987). Observations in the sea have shown that the thickness of the wave-current boundary layer in the sea can extend up to a height of 0·5 m from the bed (Slaattelid *et al.* 1987).

5.3.4. Sediment

An indication of bottom type is presented at discrete locations on Admiralty charts. The spatial distribution of surficial sediments over the UK continental shelf (UKCS) is depicted on the British Geological Survey (BGS, 1987) series of charts, 'Sea Bed Sediments and Quaternary Geology', each covering an area of 1° latitude by 2° longitude at a scale of 1:250 000. A more general coverage is provided at smaller scale on the two 1:1 000 000 sheets for the whole UKCS. Details of the original borehole logs or the grading analysis of grab samples used to construct these charts can be obtained usually from BGS. If sufficient data are not available at a site, or the data are considered unreliable (e.g. due to a known change in local conditions), it may be desirable to collect and analyse bed samples, especially in those areas where the sediment properties change rapidly spatially. Also the boundaries between the different sediment types can vary their positions seasonally.

If the bed shear stress τ_0 exceeds the threshold value τ_{cr} cohesionless sand particles begin to move across the sea bed, or muddy (cohesive) sediments start to erode (not the same definition for τ_{cr}). The value of the threshold shear stress τ_{cr} for sandy sediments depends upon the median grain size d_{50} and characteristics of the particle size grading, the mineral density of the particles and their shape, and the temperature and salinity of the water (Soulsby, 1997). The most frequently used method with which to determine the threshold is the Shields curve, an empirical expression based upon experimental data (Appendix 2, Section 5). This method uses the non-dimensional quantity θ_{cr} (also Equation 14), a convenient way of expressing the results of experiments for different ambient conditions in a unified fashion (containing ρ_s and ρ the sediment and fluid densities, g the acceleration due to gravity and the grain size d_{50}). This non-dimensional notation is also used when expressing the shear stresses τ_c and τ_w

$$\theta = \frac{\tau}{(\rho_s - \rho)gd_{50}} \tag{19}$$

The magnitude of the bed shear stress generated by a given value of wave orbital velocity or current speed is governed by the physical roughness of the sediment bed, usually denoted by the hydraulic roughness length z_0 ($= d_{50}/12$ for a flat bed of sediment; Soulsby, 1997). It should be noted that the roughness and threshold shear stress of cohesionless sediments are closely linked, i.e. whilst gravel sized material has a larger roughness length it also has a larger mass and requires a higher shear stress to mobilise it. Typical values of z_0 for muddy, sandy and gravelly sediments and mixtures of sediments are tabulated by Soulsby (1997).

When dealing with muddy sediment the threshold for erosion is usually dominated by the cohesive forces arising from electrochemical and biological sources and the degree to which the deposit becomes consolidated (or dewatered). The expression of Mitchener *et al.* (1996) relates the threshold for erosion to the bulk density of the sediment (Appendix 2, Section 7). Methods for prescribing z_0 for muddy sediments are given by Delo and

Ockenden (1992), although there is presently no consensus on how best to prescribe z_0 for mixed mud and sand sediments.

The correct specification of the sediment conditions is probably the most crucial, but difficult, part of the whole sediment mobility method.

The two alternative approaches to calculating sediment mobility are discussed below.

5.4. DESIGN WAVE–DESIGN TIDE APPROACH

In this approach one 'design' combination of the tidal current speed, water depth and wave characteristics is chosen for which the sediment mobility is calculated, the choice of environmental conditions being determined by design considerations. The sediment mobility calculations are completed using the environmental parameters as inputs to determine whether the values of τ_c or τ_w exceed τ_{cr} at a specified location, i.e. if the sediment is mobile or not. Sediment is mobilised under combinations of wave–current flow when the value of τ_{max} exceeds τ_{cr}.

5.4.1. Currents

The current related bed shear stress can be calculated from the expression

$$\tau_c = \rho C_D \bar{U}^2 \qquad (20)$$

where ρ is the water density, C_D the drag coefficient (a function of the ratio of water depth and z_0) related to $\bar{U}$ the depth averaged current speed. From Equations (17a) and (17b) $\bar{U}$ occurs at a height above the bed equal to 32% of the water depth and the tidal current speed at the surface is equal to $1{\cdot}07\bar{U}$. If necessary the influence of the meteorologically induced currents on the tidal current can be included using the procedure in Department of Energy/Health and Safety Executive (1992/93).

5.4.2. Waves

The wave related bed shear stress is calculated from

$$\tau_w = 0{\cdot}5\rho f_w U_w^2 \tag{21}$$

in which f_w is the wave friction factor, a function of the ratio A/z_0 where A is the amplitude of the orbital wave motion at the bed ($U_w T/2\pi$). For monochromatic waves, using small amplitude linear wave theory, U_w is determined as:

$$U_w = \frac{\pi H}{T} \frac{1}{\sinh(kh)} \tag{22}$$

where H is the wave height, T the wave period, k the wave number ($k = 2\pi/L$, L = wave length) and h the water depth.

In many practical applications offshore waves will have steepnesses near or above 1/20, which is the limit of validity of linear wave theory. However, with respect to the bottom orbital velocity, the predictions made by higher order wave theories are usually only marginally better than the results produced by linear wave theory, and therefore the extra computational effort invariably involved in applying higher order wave theories (such as Stokes fifth order waves) is hardly justified. In shallow water the waves become non-linear and account of this should be taken in the calculation of bottom velocity although, as discussed by Soulsby (1987), the use of linear wave theory even for waves that are about to break gives reasonable agreement with experimental observations.

5.4.3. Wave–current interaction

The bed shear stress due to the combined wave–current motion is calculated using one of the methods presented by Soulsby (1997): see Appendix 2, Section 3.

5.4.4. Sediment mobility

The mobility of the sediment can be quantified by comparing the shear stress calculated according to the method outlined above with the critical shear stress, defined from Equation (14) as:

$$\tau_{cr} = \theta_{cr}.g(\rho_s - \rho)d_{50} \tag{23}$$

where θ_{cr} is the critical Shields parameter, g is acceleration due to gravity ($9{\cdot}81\,\mathrm{m\,s^{-2}}$), ρ_s and ρ are the densities of sediment and water respectively and d_{50} is the median grain diameter. An algebraic expression depicting Shields' original experimental curve for θ_{cr} has been derived by Soulsby (1997): see Appendix 2, Section 5.

5.4.5. *Potential sediment transport rate and scour depth*

The total sediment transport rate due to waves and currents can be calculated by one of the methods described in Soulsby (1997). Present understanding leads to the conclusion that the most important waves for long-term sand transport predictions offshore are those with about 10% exceedance. Convenient maps of the 10% exceedance wave height are presented by Draper (1991). The sediment transport methods can also be applied to estimate the transport of stone by waves and currents from an area of rip-rap.

The potential equilibrium scour depth is given by the ratio of bed shear stresses τ_0/τ_{cr}, as discussed in Chapter 7, although the scour depth actually attained is also related to the sediment transport rate which determines T^*, the timescale for the development of scour.

5.5. PROBABILISTIC APPROACH

A more thorough approach is to consider the likelihood of sediment being mobilised from around the completed structure rather than to perform calculations for just one or more design conditions. The calculations required are more onerous than in the design method because the probability distributions of waves and currents are combined to produce a probability distribution for τ_{max}.

Previously, Katsui and Bijker (1986) and Myrhaug (1995) have proposed methods with which to predict the probability distribution of the peak shear stress due to waves, assuming stationary Gaussian narrow-banded distributions for the bed orbital velocity and surface wave elevation respectively.

The probabilistic approach proposed here uses information on the long-term distributions of waves and currents (measured or synthesised over a duration of at least one year) as inputs to derive a probability distribution for the bed shear-stress. Once this distribution has been produced it is then relatively simple to define the probability of sediment becoming mobile both on the undisturbed bed and in the vicinity of the structure.

The amount of information available to make the calculations varies but would typically consist of (1) wave data (joint frequency distribution of wave height and period $p(H_s, T_z)$) and (2) currents data (frequency distribution of current speed $p(U_c)$). The principle of the method is as follows.

A distribution of the bed peak orbital velocity, $p(U_w)$, is derived from the omnidirectional distribution $p(H_s, T_z)$. If omnidirectionality is also assumed for the currents as for the waves, a joint frequency distribution for waves and currents, $p(U_w, U_c)$, is estimated assuming that the probabilities $p(U_w)$ and $p(U_c)$ are independent, i.e. $p(U_w, U_c) = p(U_w)p(U_c)$. In many applications the currents will be predominantly tidal and the waves will be locally generated or swell thus rendering the independency assumption reasonable.

Of course, this assumption may not be valid when both currents and waves are wind generated, i.e. under storm surge conditions in shallow water (e.g. CIRIA, 1996), nor is it strictly valid in the shallow nearshore region where the largest currents may be wave-induced and the wave characteristics cannot be treated always with linear wave theory. Under both these conditions a more rigorous approach to deriving $p(U_w, U_c)$ is required.

Given a distribution for $p(U_w, U_c)$ the corresponding distribution of wave–current generated bed shear stress, $p(\tau_{max})$, can be calculated. Comparing $p(\tau_{max})$ with the threshold shear stress for sediment motion will indicate the potential for mobility and attach a probability to it.

In the following section the different steps of the probabilistic approach are dealt with in more detail. All distributions are discrete and omnidirectionality is assumed. The approach can be applied for waves or currents alone as well as for the combined wave–current case.

5.5.1. Distribution of current speeds

If at a location the current magnitude is non-zero and the current speeds vary appreciably, a distribution of the hourly current speeds over one year can be obtained. For most practical purposes the currents will be predominantly tidal and, thus, tidal information can be used to construct $p(U_c)$. For most situations a discrete interval of $0{\cdot}1\,\mathrm{m\,s^{-1}}$ for the distribution of current speed should suffice.

5.5.2. Distribution of bottom orbital velocity amplitude

For irregular waves the determination of U_w is complex as the wave climate consists of simultaneous distributions of H and T rather than discrete values, for example as determined every 3 hours from a Waverider buoy. The wave height and period are identified by the significant wave height, H_s, and zero-crossing period, T_z. In most cases it is sufficient to discretise the data into 1 second T_z bands and $0{\cdot}5$ m H_s bands.

For a random sea the bottom velocity spectrum $S_u(\omega)$ is determined as:

$$S_u(\omega) = H(\omega)^2 S_\eta(\omega) \tag{24}$$

in which ω is the radian frequency ($\omega = 2\pi/T$), $H(\omega)$ is the frequency response function and $S_\eta(\omega)$ is the surface elevation spectrum (usually JONSWAP or Pierson–Moskowitz) corresponding to given values of H_s and T_z. It should be noted that for the standard JONSWAP surface elevation spectrum the ratio T_z/T_p is $0{\cdot}781$ and for the Pierson–Moskowitz spectrum $0{\cdot}710$ (Soulsby, 1987), the peaks in the corresponding elevation and velocity spectra for the former case are sufficiently close for practical purposes. Assuming linear wave theory (Equation (22)), the frequency response function is given as:

$$H(\omega) = \frac{\omega}{\sinh(kh)} \tag{25}$$

where h is the water depth. The root-mean-square (rms) value of the bottom orbital velocity, U_{rms}, is determined from:

$$U_{rms}^2 = \int_0^\infty S_u(\omega)\, d\omega \tag{26}$$

The procedure is repeated for each discrete combination of H_s and T_z values taken from the joint frequency distribution. U_w can now be found from the relation:

$$U_w = \sqrt{2}\, U_{rms} \tag{27}$$

5.5.3. *Shear stress distribution*

The probability distribution of the maximum shear stress is assumed to be

$$p(\tau(U_{c,i},\ U_{w,j})) = p(U_{c,i}, U_{w,j}),\ i = 1, \alpha;\ j = 1, \beta \tag{28}$$

where α is the number of discrete current velocity intervals and β is the number of discrete wave orbital velocity intervals, for which the bed shear stress for $\tau(U_c,\ U_w)$ has been calculated according to the procedure in Appendix 2.

The cumulative exceedance probability function of τ_{max} exceeding a specified value τ^1 is defined as:

$$P(\tau_{max}^1) = \mathrm{Prob}[\tau(U_c, U_w) \geq \tau^1] = \sum_{(i^1, j^1)} p(U_{c,i}, U_{w,j}) \tag{29}$$

where (i^1, j^1) is all the combinations of (i, j) for which $\tau(U_{c,i}, U_{w,j}) \geq \tau^1$.

The probability for scour is determined as the part of the distribution $P(\tau_{max}^1) \geq \tau_{cr}/M$.

It should be noted that the above is in fact a quasi-probabilistic approach but has the advantage that it is generally applicable, simple to use and will be sufficient for most applications. The methods and assumptions of Katsui and Bijker (1986) and Myrhaug (1995) both allow a more rigorous probability distribution for the wave shear stress to be derived, but for the case where there is no current. As has been discussed, the addition of a current can, in many circumstances, substantially increase the value of the maximum bed shear stress from the wave alone case.

As a refinement of the above, for instance if information is required to help prescribe the orientation of a structure, it might be necessary to take into consideration the directional information of waves and currents. In this case one of two procedures can be adopted but they are more complicated and laborious:

- discretise the compass rose into a finite number of sectors (e.g. $12 \times 30°$) and then determine $p(\tau_{max})$ from the wave and current inputs corresponding to each sector individually, or
- calculate simultaneously the values of τ_{max} for every combination of the wave orbital velocity and direction and the current speed and direction, and then discretise the results into sectors. The shear stress in this case can be predicted directly using one of the wave–current interaction formulae including the influence of wave–current direction ϕ parameterised in Soulsby (1997).

If the variation in water depth at the site is significant then the calculations should be run for a range of water levels.

5.5.4. *Sediment transport rates*

If the bed is generally mobile it can be useful to calculate the probability distribution of sediment transport rates. This can be approximated from the product of the combined probability of occurrence of waves and currents, and the transport rate for that combination:

$$p(q)_{ij} = p(U_{c,i}, U_{w,j})q_{ij}(U_{c,i}, U_{w,j}), \quad i = 1, \alpha; \; j = 1, \beta \qquad (30)$$

The long-term average of the (gross) sediment transport rate for combined waves and currents is calculated from Equation (30) using the method presented by Soulsby (1997).

Katsui and Bijker (1986) proposed a method for estimating the expected transport rate of bed material under random wave action and applied it to a worked example of the displacement of rip-rap. A similar approach for the combined wave current case could be devised.

As the potential equilibrium scour depth S_e is given by the ratio of bed shear stresses τ_0/τ_{cr} and a distribution of τ_0 values is available from this method, the likelihood of reaching a specified scour depth from an initially flat bed can be estimated. However, to make the method, more realistic the duration of events of a given τ_0 must be taken into account to enable the joint influence of $S_e(\tau)$ and $T^*(\tau)$ to be represented. Of course the exact timing of when a scour depth of a given value first occurs relies on the chronology of wave and current forcing, i.e. shear stress input.

5.6. CONCLUSIONS

A methodology for quantifying the degree of sediment mobility and hence the potential for scour at marine structures has been presented. The methods described can be programmed on a personal computer to predict whether scour occurs at a single location either under specified design wave–design tide conditions or for a specified wave–current climate.

Once the calculations have been performed for a single location it is a simple matter to perform them at each of a whole array of grid points. With this information a map of the scour potential in the vicinity of a planned or existing structure, or over a larger area of the sea bed in which a suitable location for placing a structure is being sought, can be produced.

The methods are versatile and are not restricted to solving sea bed–structure interaction problems; for example, they have recently been used to assist in the choice of suitable areas of sea bed for gravel extraction by allowing the number of hours per year for which the gravel is mobile to be calculated and plotted as a contour map (Figure 13). This figure was derived as part of a research study funded by the Crown Estate and SCOPAC (HR Wallingford, 1993).

5.7. EXAMPLE OF CALCULATING SEDIMENT MOBILITY

The probabilistic approach is illustrated below with a hypothetical example.

An engineer is required to examine the need for scour protection at a slender vertical pile founded within a uniform depth nearshore location in 10 m of water (mean depth) and exposed to waves and tidal currents. To help in assessing the situation it is necessary to quantify the likelihood for scour occurring under the prevailing hydraulic climate.

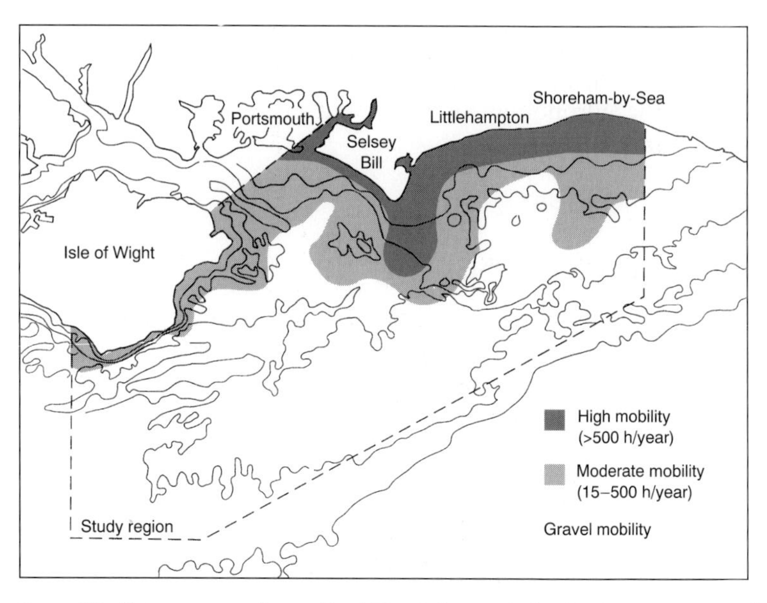

Figure 13. Contour map of gravel mobility (after work for the Crown Estate and SCOPAC, HR Wallingford, 1993)

The information available is as follows:

- Current speeds: Current speed data (Figure 14), derived from a numerical model validated against field measurements.
- Wave climate: a discrete H_s–T_z scatter plot derived from one year's worth of non-directional Waverider buoy measurements, shown in Figure 15.
- The mean water depth is 10 m.
- The local sediment characteristics are $d_{50} = 0{\cdot}010$ m (shingle), $\rho_s = 2650\,\text{kg}\,\text{m}^{-3}$.
- The average water characteristics are viscosity, $\nu = 1{\cdot}2\times10^{-6}$ $\text{m}^2\,\text{s}^{-1}$ and density, $\rho = 1026\,\text{kg}\,\text{m}^{-3}$.
- The angle between the current and wave direction ϕ has been set to 45° as an average.
- The shear stress amplification factor for the pile is $M = 4$ (see Table 2).

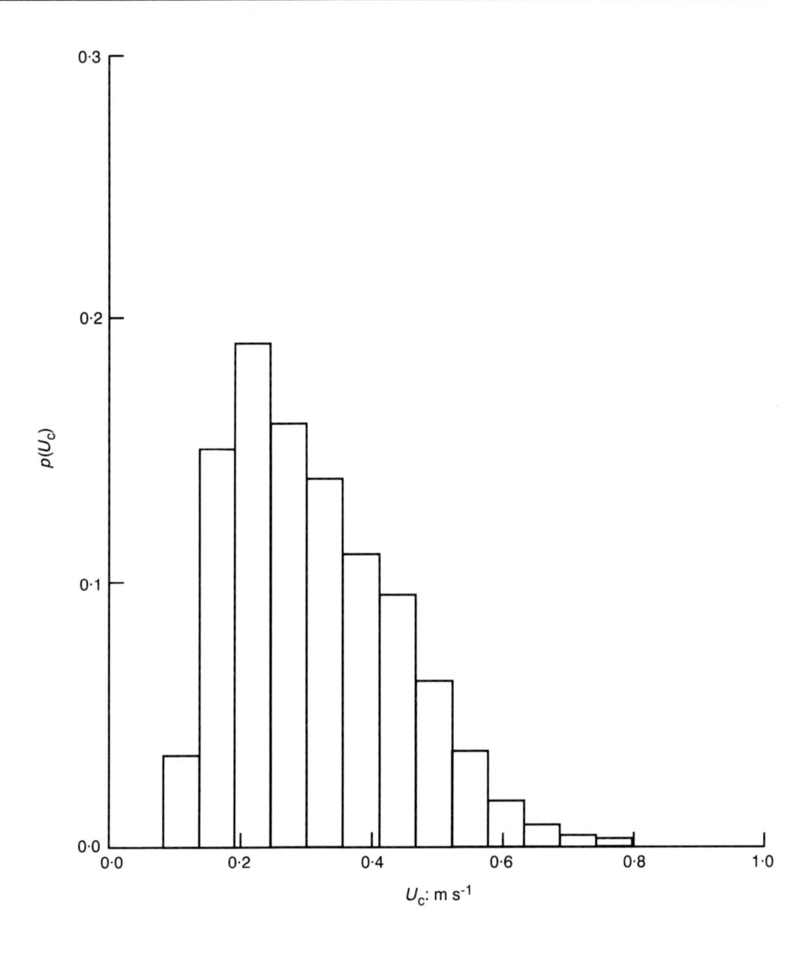

Figure 14. Discrete distribution of current speeds

The procedure for determining the bed shear stress and the exceedance probability for sediment motion is as follows, with the results of the calculations depicted graphically.

Step 1: Determine the discrete distribution of the current speed $p(U_c)$ over a period of one year. The resulting distribution is shown in Figure 14.

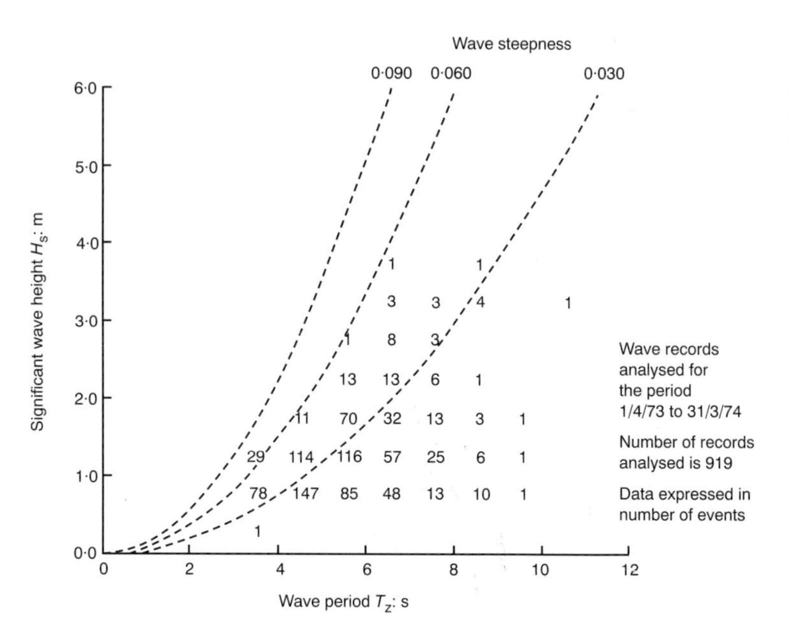

Figure 15. H_s : T_z distribution of one year's Waverider data

Step 2: Use the procedure in Section 5.5.2 to determine the annual distribution of $p(U_w)$ from the H_s : T_z data (Figure 15). The resulting distribution of $p(U_w)$ is shown in Figure 16.

Step 3: Determine the joint probability density function for wave orbital velocity and current speed. This is done by multiplying $p(U_w)$ with $p(U_c)$ for each discrete combination of (U_w, U_c). The result is shown in Figure 17.

Step 4: Calculate mean and maximum wave–current generated bed shear stresses using the procedure in Appendix 2. The resulting shear stresses for the given wave and current intervals are shown in Figure 18. On the basis of $p(U_w, U_c)$ and the calculated shear stresses, the cumulative exceedance probability distribution, $P(\tau_{max})$ can be determined according to Equation (29). A graph of $P(\tau_{max})$ is shown in Figure 19.

Step 5: Calculate τ_{cr} using Equation (23) with $\theta_{cr} = 0{\cdot}055$ for $d_{50} \geq 0{\cdot}01$ m: $\tau_{cr} = 8{\cdot}8\,\mathrm{N\,m^{-2}}$.

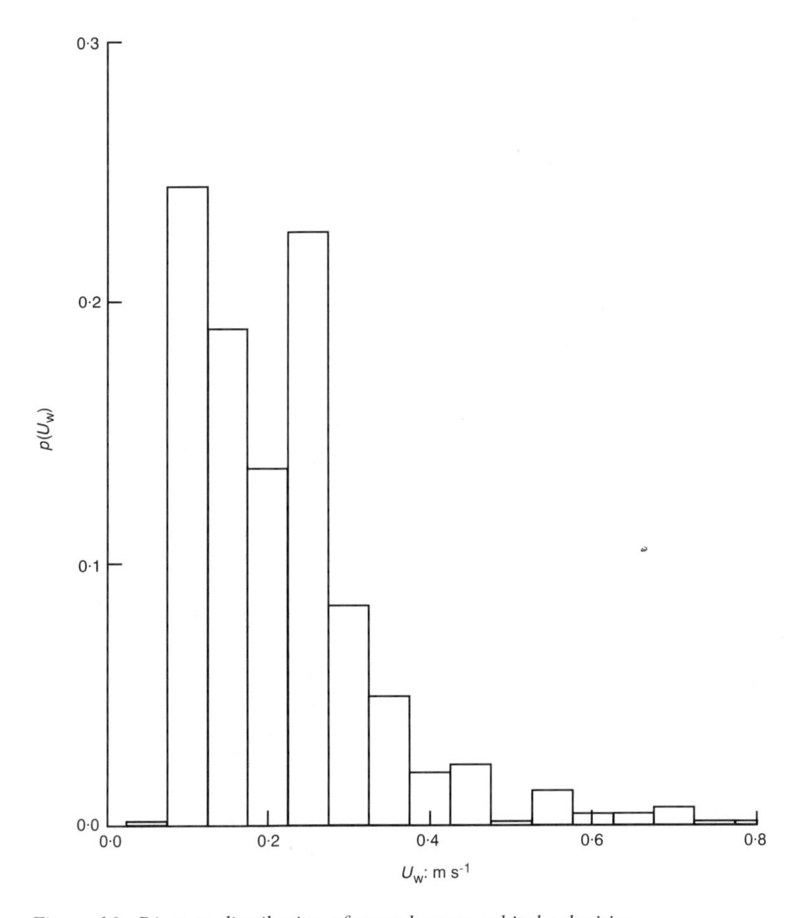

Figure 16. Discrete distribution of wave bottom orbital velocities

Step 6: Now the value of τ_{cr} can be compared to $P(\tau_{max})$. Under the prevailing conditions the bed material is not expected to be mobile for a significant amount of the time.

Step 7: The probability of scour occurring around the pile corresponds to the probability value at which $\tau_{cr}/M = 8{\cdot}8/4 = 2{\cdot}2$ N m^{-2}. For this example, the material adjacent to the base of the pile is expected to be mobile for about 10% of the time.

A decision about the need for bed protection can be made using this information. The options available for bed protection including a sizing method for stone are discussed in Chapter 6.

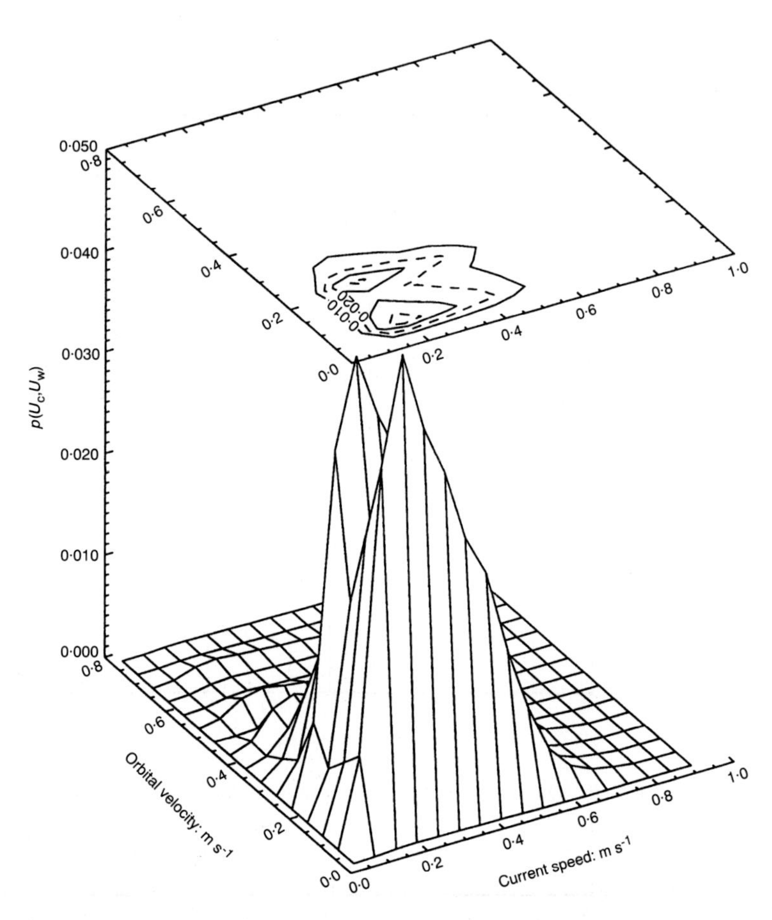

Figure 17. Contour plot of joint probability of wave orbital velocity and current speed

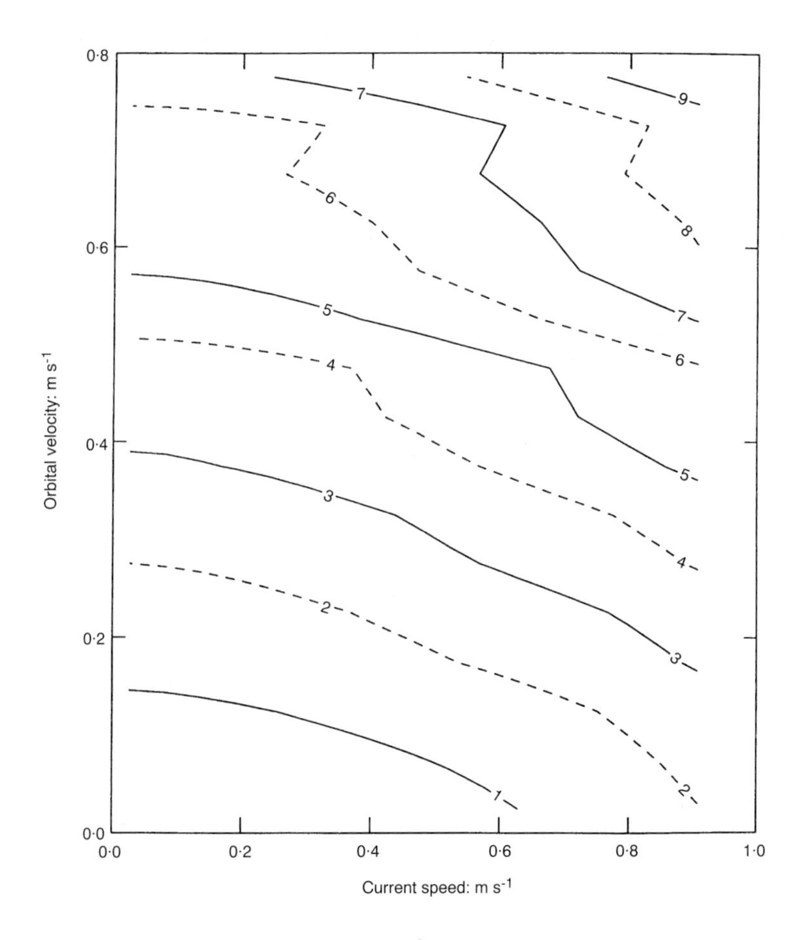

Figure 18. Contoured plot of τ_{max} (Nm^{-2}) for combined waves and currents

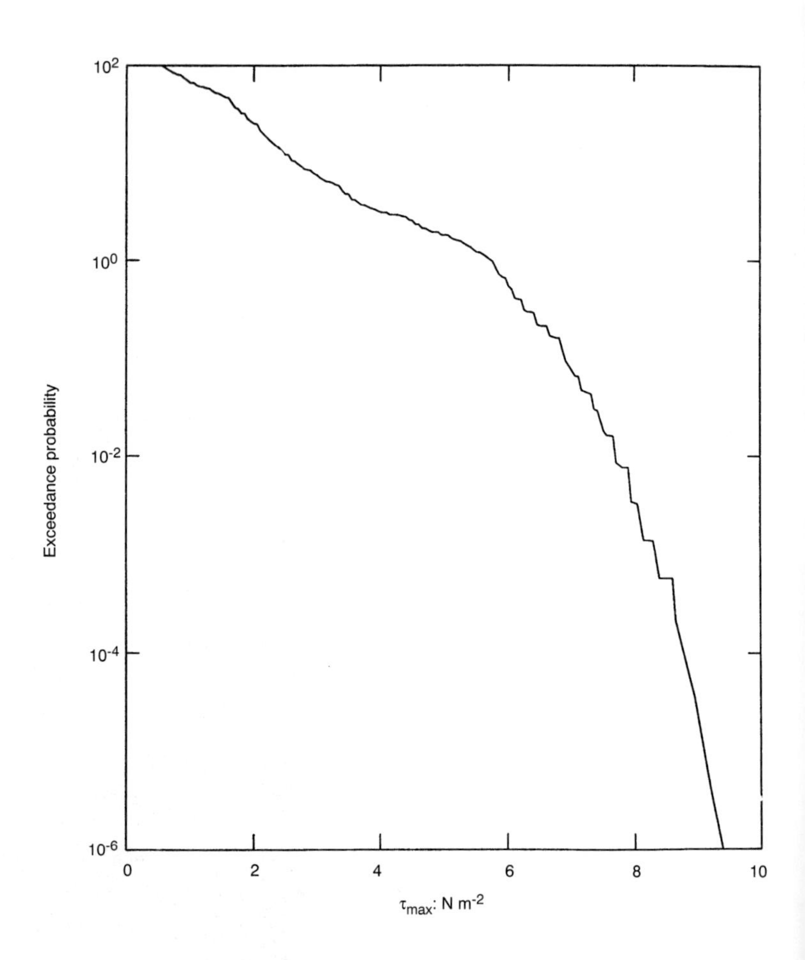

Figure 19. Exceedance probability for τ_{max}

Preventive or remedial measures for scour

6. Preventive or remedial measures for scour

The majority of the case studies and examples referred to in Chapter 7 of this report deal with the scour development at unprotected structures. In practice the need for scour protection should be considered at an early stage in the design process as the retrofit of scour protection or the implementation of remedial works is generally costly, and may have consequences for the rest of the works. The occurrence and impact of scour during the construction period must be planned for as the works required to counteract scour and bring the substrate back to grade might lead to a significant overrun in construction costs (e.g. Hales, 1980a). The cost of post-project completion remedial work is often borne by maintenance budgets and hence it is isolated from the original construction costs.

The construction industry codes specify that an adequate assessment of scour be included at the design stage and, if necessary, that appropriate preventive measures be provided. The notes appropriate to offshore structures (e.g. Department of Energy/Health and Safety Executive, 1992/93) indicate methods for consideration at certain types of installation; i.e. skirts underneath gravity base structures, grout or sand bags to underpin pipelines with a mattress or gravel dump covering.

Dahlberg (1981) has reviewed the different methods of scour protection that have been employed for gravity structures in the North Sea:

- perforated breakwater wall
- extended base slab
- well graded gravel over a width of minimum 10% of the base diameter along the periphery
- artificial seaweed (carpets or curtains)

- sand/gravel bags.

Experiences with using rock dump, artificial seaweed and mattresses in the North Sea are described by Knott (1988).

Scour protection measurements are often required at coastal structures (Oumeraci, 1994a) where the influence of shallow water, breaking waves and longshore currents can become significant. Many different types of structure will need to be protected including monopile installations and multiple piled jetties, pipelines, sea walls and breakwaters. As in the offshore situation rock dumping is often used to provide a protective layer against scour. The physical properties of rock and engineering guidelines for its use in and around coastal structures are described in the comprehensive manual produced by CIRIA/CUR (1991). Examples of scour protection and scour stabilisation works used for hydraulic structures such as barriers are discussed by Dietz (1995), Heibaum (1995) and Hoffmans and Verheij (1997).

Obviously the optimum approach to scour prevention is to minimise the scour potential at the structure to an acceptable level, through careful and informed design, and then to consider the need for a scour protection system to provide an additional level of confidence. A method for assessing the likelihood of scour has been suggested in Chapter 5. Complementary design approaches will include careful siting of the structure, based upon adequate site investigation and environmental data, provision of secondary structures to reduce the influence of scour at the primary structure or careful preparation of the substrate prior to commencing the installation or construction work. The importance of correctly preparing the bed ahead of breakwater construction is demonstrated by Hales and Houston (1983). At offshore locations scour protection measures have to be installed quickly to reduce the impact of scour (Watson, 1979), with attention first being given to the corners if an angular structure is being used. For example in the southern North Sea, scouring can take place around structures within a period of 2 weeks following installation.

The methods currently available for protection against scour, wave-current loading or other impacts in the marine environment, or for increasing on-bottom stability are:

- protective apron
- rock dumping for bed stabilisation

- mattresses
- trenching (pipelines) or increasing structure embedment
- sand bags or grout filled bags, for support or bed stabilisation
- concrete saddles
- anchoring, with strops and soil anchors
- flow energy reduction devices, including artificial seaweed
- soil improvement, to increase bearing capacity and reduce scour potential.

The use of these is listed in Table 5 and discussed below.

6.1. PROTECTIVE APRONS

In this section the experience of minimising scour at piled structures by the use of solid collars installed around the base of the pile is reviewed. The aim of these types of device is to protect the bed from the scouring influence of the downflow at the pile and the associated vortex action around the base of the pile.

Table 5. Preventive or remedial measures for scour: summary of usage

	Structure					
Method	Piled structures	Pipelines	Large volume structures	Sea walls	Break-waters	Jack-up platforms
Protective apron	●		●	●	●	
Rock dumping	●	●	●	●	●	●
Mattresses		●	●	●	●	●
Trenching or embedment		●	●			●
Sand/grout bags	●	●	●			●
Flow energy reduction	●	●	●	●	●	●
Soil improvement	●		●	●	●	●

Based on the investigations of the flow structure by several workers it is found that the scour collar needs to extend a distance of at least one cylinder radius from the outer wall of the pile to prevent the flow from impinging on the bed. Hjorth (1975) undertook laboratory tests with a collar of annular width equal to $D/2$ and still measured values of the bed shear stress equal to 3 times the ambient value on the unprotected bed just outside the collar (i.e. $M = 3$). This indicates that a collar of this size would not be completely effective in protecting the bed from scour. However, the tests of Rance (1980) indicate a protective apron extending to $0{\cdot}5D$ from the wall of a circular structure, and $1D$ from the wall of an angular structure, might in practice offer enough protection against scour.

A rule of thumb adopted by river engineers (e.g. Breusers and Raudkivi, 1991) is that the width of the protection measured from the pier should be three or four times the projected width of the pier.

The data on the performance of collars relate to the clear water scour condition, i.e. no sediment transport upstream of the pile. The performance with general sediment motion has not been investigated and the effectiveness of a collar as a bed protection mechanism remains to be studied for this situation.

To be effective the collar must remain flush with the bed surface at all times otherwise flow can pass underneath and accelerated erosion will take place. In addition the collar must not be so thick that it causes an obstruction to the flow and promotes scour.

Problems with gaps forming under fixed collars allowing scour due to underflow to proceed has led to an interest in the use of collars which slide down the pile as the bed is lowered, thus offering protection at the lowest achieved bed level. Another approach is to use articulated (hinged) plates which can also account for a lowering in the general bed level around the pile. The performance of hinged concrete slabs was found to be satisfactory as a scour protection measure around the 'Nordsee' (gravity base type) research platform (Maidl and Schiller, 1979; Maidl and Stein, 1981), bottom diameter 75 m installed in 30 m of water. Some slabs failed but this was attributed to poor hinge design.

In the southern North Sea, where the shallow water results in significant wave activity at the bed and high sediment transport rates, the use of fixed inflexible concrete aprons is not recommended because they restrict the dissipation of excess pore water pressures and can be quickly undermined by scour (Watson, 1979). Replacement of this type of bed protection is difficult and costly.

6.2. ROCK AND GRAVEL DUMPING

An alternative approach to the fixed apron or collar is to install a protective layer of rip-rap (stone or gravel) around the base of the pile. Techniques for the placement of rock from surface vessels are reviewed by Herbich *et al.* (1984) including:

- from a side-dumping barge or vessel with individual stones falling to the sea bed
- from a split-hopper barge as one big mass
- from a barge through a pipe to reduce the fall velocity of the rock and improve placement accuracy.

The choice of method depends upon the total amount and size of material to be used and the specific location for which the material is required. For coastal structures other forms of placement may also be considered (CIRIA/CUR, 1991).

The material used to form the protective layer must offer sufficient resistance to withstand the flow induced forces (enhanced shear stress, vortex action). The stability of the local sea bed material can be calculated based upon knowledge of the local flow field around the pile (Chapter 2). Calculating the bottom shear stress becomes complicated owing to the fact that the shear stress acting on the protective layer is governed not by the characteristics of the ambient bed material but by the size of the rip-rap material, which may not be known *a priori*. An initial estimate of the size of material that will be stable under design conditions can be made for the depth-averaged current velocity and/or the bottom orbital velocity due to the wave action by using the approach of Soulsby (1997).

Recent research on the stability of bed material under the action of waves and currents has been reported by Soulsby and Whitehouse (1997). To calculate the grain diameter d_{cr} which is just immobile for a given flow Figure 20 may be used for currents and Figure 21 for waves (from Soulsby, 1997).

Based upon the data for threshold shear stress yielding Equation (75) in Appendix 2, the value of θ_{cr} becomes approximately constant at 0·055 for large grain sizes (dimensionless

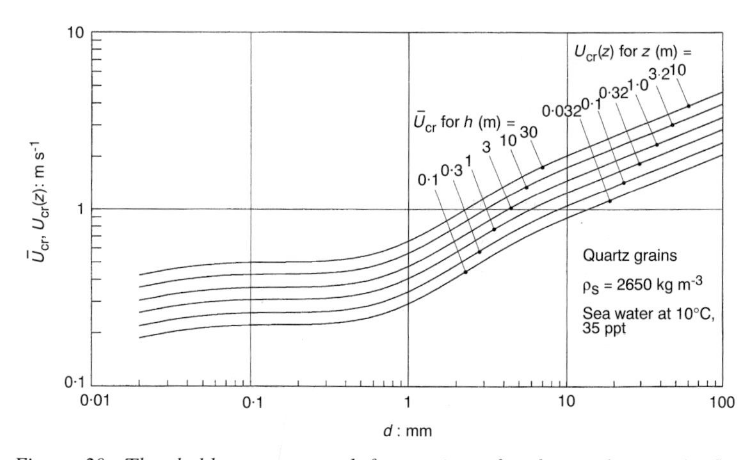

Figure 20. Threshold current speed for motion of sediment by steady flows (reproduced from Soulsby, 1997)

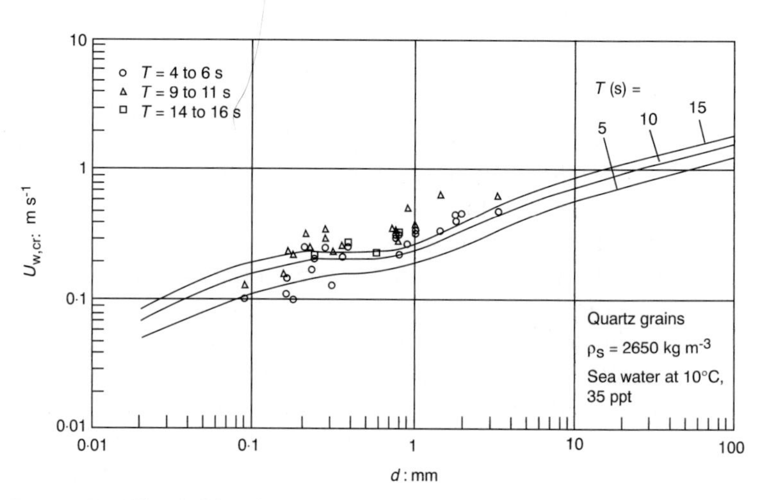

Figure 21. Threshold orbital velocity for motion of sediment by waves (reproduced from Soulsby, 1997)

$D_* > 200$; see Equation (74) in Appendix 2 for definition of D_*). For quartz density grain sizes larger than 10 mm in sea water Soulsby (1997) suggests the following equations to calculate the threshold grain diameter which is just immobile for given flow conditions. These equations relate to the lines on Figures 20 and 21 for material larger than 10 mm. Strictly the expressions are validated against data up to a grain size of about 40 mm in sea water ($D_* \sim 800$).

The formula for steady flow is

$$d_{cr} = \frac{0{\cdot}250\, \bar{U}^{2{\cdot}8}}{h^{0{\cdot}4}\, [g(s-1)]^{1{\cdot}4}} \quad \text{for } d > 10\,\text{mm} \tag{31}$$

and the formula for waves is

$$d_{cr} = \frac{97{\cdot}9\, U_w^{3{\cdot}08}}{T^{1{\cdot}08}\, [g(s-1)]^{2{\cdot}08}}, \quad \text{for } d > 10\,\text{mm} \tag{32}$$

where

d_{cr} = grain diameter which is just immobile for a given flow
$\bar{U}$ = depth-averaged current speed
h = water depth
U_w = wave orbital velocity amplitude at sea bed
T = period of water waves
g = acceleration due to gravity
s = ratio of densities of grain and water

An iterative approach based on calculating the bed shear stress can be used to refine these estimates. Verification can be provided by physical model tests or prototype investigations.

The stability of the rock material which has a sloping surface can be calculated using the conventional threshold approach for horizontal beds with a correction made for the angle of inclination of the bed to the horizontal using the equation:

$$\frac{\tau_{\beta cr}}{\tau_{cr}} = \frac{\sin(\phi_i + \beta)}{\sin \phi_i} \tag{33}$$

where β is the local bedslope in the along flow direction (degrees) and ϕ_i is the angle of repose of the rockfill material. β is defined as positive when the flow is up the slope and negative for the downslope case. Assuming τ is proportional to velocity squared, the square root of Equation (33) will give the slope correction factor in terms of velocity. Equation (33) has been validated against data by Whitehouse (1995).

In stability investigations both the shear stress amplification M due to the presence of the structure and effects related to the disparity in size of ambient bed material and rip-rap must be considered. The transition from the sea bed to the rip-rap layer presents a transition in bed roughness which produces an overshoot in the bed shear stress downstream of this transition in a steady current (Soulsby, 1983) as well as under waves (Sumer *et al.*, 1991); under wave flows there is also a steady streaming induced in the bottom boundary layer by the differential roughness. The shear stress adjusts quite quickly after the physical transition point but the bed roughness length takes much longer to adjust in the streamwise direction. The combined influence of the overshoot in shear stress and insufficient coverage to protect the bed from the overall increase in shear stress due to the presence of the pile could lead to an increased susceptibility for damage at the edge of the rip-rap layer. The bed protection must extend far enough to remove the influence of these edge effects from the primary structure and the thickness of the blanket needs to be profiled carefully to reduce scouring at the edges due to over-thickness and disturbance to the flow. The laying of a thick mattress could exacerbate the scour development around the area of the blanket leading to a greater chance of failure of the scour protection. In laboratory tests to examine the performance under wave/current action of scour protection material around large bridge caissons, Hebsgaard *et al.* (1994) observed scour depths of 3 m in the sea bed sediments just downstream of the protection layer. The width of the protection layer was 15 m.

Thus the bed protection design must take into account the possible interaction between adjacent piles or structures and the downstream effects caused by the structure (and possibly the scour protection). Bed stabilisation using gravel dumping either pre-installation or post-scouring has proved to be effective against scour at jack-up rig footings (Sweeney *et al.*, 1988; Lyons and Willson, 1986). Laboratory tests (Sweeney *et al.*, 1988) showed that backfilling of the scour hole with rock could successfully stabilise the scour hole but not necessarily prevent further spud can settlement. Further tests showed that a 2 m wide gravel apron with a filter layer of stone considerably reduced settlement due to scouring. This apron could be extended to include the area around a jack-up spud can and jacket foundation in close proximity.

The laying of the rip-rap must follow certain rules to ensure that the ambient bed material is not lost through the pore spaces and that gradients in the pore pressure can be dissipated. The rip-rap design is recommended to follow a reverse filter criterion of at least two layers' thickness with the specified bed protection material forming the topmost layer. The placement of rocks directly onto a sand bed could result in them scouring and settling due to the action of waves and currents (e.g. Nago and Maeno, 1995).

The correct filter performance is achieved by using a well graded mixture which allows the voids between the larger stones to be filled by the smaller sizes in the mixture, thus producing a layer with reduced porosity but high internal support. The gradings of the first layer (and successive ones) are specified as follows:

$$d_{15,\mathrm{f}} \leq 5d_{85,\mathrm{b}} \tag{34a}$$

$$4 \leq \frac{d_{15,\mathrm{f}}}{d_{15,\mathrm{b}}} \leq 20 \tag{34b}$$

$$\frac{d_{50,\mathrm{f}}}{d_{50,\mathrm{b}}} \leq 25 \tag{34c}$$

where d_{15} is the size of bed material for which 15% by weight is smaller etc. and the subscripts refer to the filter f and the bed material b respectively. These guidelines should be followed when placing multiple graded layers and when using rockfill in other situations.

In certain circumstances the rip-rap might be laid on a geotextile membrane rather than a rock filter, with a layer of smaller quarry-run stone to prevent the material from being damaged by the larger rocks. When using geotextiles care must be taken to ensure that the membrane does not become clogged by fines thus reducing the potential dissipation of pore water pressures.

De Wolf *et al.* (1994) describe the steps involved in the design and installation of monopile research platforms off the Belgian coast, including one in 14 m of water on top of the Westhinder bank, medium sand (d_{50} = 0·280 mm). The necessary erosion protection to cope with the predicted scour depths of 4 to 6·3 m extended 22 m from edge to edge with the 2 m diameter pile in the centre. Two layers were used, a gravel layer 1 m thick covered by a 1 m thick layer of quarry run stone (2–300 kg).

Bed protection for the pile clusters used to support offshore structures or bridge structures with a pile cap can be achieved using techniques appropriate for single piles, but the projected width of the cluster must be used in calculations. Vitall *et al.* (1994) measured a reduction in scour at a 3-pile group due to the addition of a collar equal to 2 times the diameter of the inscribed circle formed by the outer edge of the pile group.

Hales (1980a) recommends the laying of an underlayer ahead of construction works in coastal regions to reduce the disruption caused by scour. Gökçe *et al.* (1994) demonstrate that a rock apron extending a distance of 3 to 4 times the width of a vertical breakwater from the breakwater face provides reasonably complete protection against wave induced scour (see Figure 40).

In the same way that a river bed consisting of a sand and gravel mixture becomes armoured with time as the finer grains are winnowed out by the less frequent higher magnitude (flow) events, the upper layers of a graded gravel armour layer will be modified with time. Rock dump has been used in the North Sea to cover the Ekofisk–Emden pipeline (Roelofsen, 1980) and it is observed that the finer material will erode from the top layers of the rock protection at conditions less than the ultimate design condition. Providing the appropriate rock grading is used the upper layer of the bed protection scheme is likely to become armoured and perhaps somewhat more resistant with time.

Another material considered for use in bed protection schemes is steel mill slag, which is cheap, reasonably durable in sea water and has a high specific gravity (e.g. Posey and Sybert, 1961). Posey and Sybert (1961) specified a 3 layer steel mill slag protection mat, using filter criteria (Equation (34)), as a remedial measure for dishpan scour under a jacket structure founded on a very fine sand bed (d_{50} = 0·085 mm) in the Gulf of Mexico. This appeared to perform satisfactorily. Steel slag has been used in the North Sea to stabilise the sea bed before any problems arise or as a material for remedial works (Van Dijk, 1980; Dahlberg, 1981).

In the nearshore region the effect of breaking waves on the stability of rock dump and mattresses has not been quantified and requires further research to provide appropriate guidelines. Hales and Houston (1983) report laboratory tests to determine the stability of the underlayer material used in a 0·6 m thick blanket beneath rubble mound structures. They concluded that the material was least stable where the waves were breaking and plunging right on top of the material at the toe of the structure; their design equation was for this condition. The influence of high turbulence levels on the stability of rip-rap and concrete block mattresses in steady flow has been investigated at HR Wallingford by Escarameia and May (1995) who produced design equations to take account of varying levels of turbulence in the flow.

6.3. MATTRESSES

Prefabricated mattresses are often used within bed protection or preparation schemes as they can be installed in a controlled manner, there is usually less control over the rock dumping method. Mattresses are often used to afford protection to pipelines but can be adopted for other sea bed structures (Maidl and Stein, 1981). A major advantage of mattresses is that they are flexible and can be laid to follow the local bed contours.

The different types of protective mattress are described by Herbich *et al.* (1984):

- fascine mattress – a synthetic filter fabric strengthened with synthetic or natural fascines, usually overlain by rock dump material
- block mattress – a continuous array of concrete blocks held together by cables and laid on the sea bed or individual blocks held in a pattern on the sea bed by synthetic nails
- cell mattress – mesh baskets filled with sand or gravel, large rocks in large wire mesh also called gabion baskets
- 'Colcrete' mattress – the mesh baskets of the cell mattress are filled with underwater concrete instead of ballast
- stone asphalt mattress – a synthetic filter fabric ballasted with a stone and asphalt mixture

- ballast mattress – a heavy synthetic fibre woven mattress is double folded at both sides and filled with sand or gravel.

Whilst the rock rip-rap is held in place due to its own weight and resistance between the rock and the underlying layers, the mattresses are often held in place by the use of soil pins or anchors. The resistance to pull out presented by the soil fixings, or the tensile strength of the material joining the mattress to the anchor, is designed to resist the uplift and drag due to hydrodynamic forces. Poorly designed fixings have historically been the most common cause of failure of these types of protection devices.

Gravity base structures are usually fitted with a skirt to provide additional stability when installed and to prevent scour channels which form at its perimeter from penetrating underneath the structure. Steps are often taken to fill any unevenness in the bed beneath the structure and this provides additional protection from scour. The satisfactory performance of mattresses as scour protection devices around concrete gravity base research platform 'Nordsee' has been reported by Maidl and Schiller (1979) and Maidl and Stein (1981). The use of Colcrete mattresses provided excellent protection against scour and they were considered to have a service life at this North Sea location of 15 years, as opposed to 1 to 2 years for the sandbag clusters.

Wilson and Abel (1973) reported on the development of the bed around a semi-submersible platform on a sandy substrate in 27 m of water. The structure foundations comprised three 24 m diameter, 8 m high circular pontoons. Laboratory and field investigations indicated that scour would occur and that this could be prevented by the installation of a scour apron comprising a radial array of mats (3·6 m wide, 15·2 m long) formed from inexpensive nylon mesh held down by steel pins and sandbags.

Mattresses have been used as part of a composite gravel fill–mattress–rock layer protection material in the southern North Sea to cover pipelines (Van Dijk, 1980; Dahlberg, 1981).

Reddy *et al.* (1990) discussed the main elements of a pipeline scour protection system for a concrete coated iron pipe in the coastal zone. The main elements of the system were a 4·2 to 4·8 m wide precast concrete flexible mattress overlaying a geotextile underlayer with anchors to hold the layers down.

This system was installed over the trenched and backfilled pipeline to prevent scouring and exposure of the pipeline.

Pipeline protection against damage from impacts due to anchors or falling objects can be provided with concrete saddles although the possible scour interaction with the sea bed should be assessed prior to their use.

6.4. TRENCHING (PIPELINES) OR INCREASING STRUCTURE EMBEDMENT

Increasing the embedment of a structure in the sea bed provides sheltering from wave and current forces, increases the stability due to a greater soil–structure interaction and also affords an increased margin of safety against undermining by scour.

The scour protection afforded to a pipeline by trenching has been demonstrated by Fredsøe (1978) and the added stability plus the reduction in hydrodynamic loads on the pipe itself by Wilkinson *et al.* (1988). Hjorth (1975) demonstrated that the flow field around a half buried pipeline was little different to the ambient case and Kjeldsen *et al.* (1974) measured a reduction in scour depth for a half buried pipe compared to the case where the bottom of the pipe rests at the initial bed level (see Figure 29, top panel). When pipelines are installed in harsh environmental conditions it may be necessary to stabilise them with anchors (Jinsi, 1986) if trenching or burial is not feasible.

Laying of pipelines through areas of mobile sandwaves may require the sea bed to be ploughed flat prior to installation; the sandwaves are then likely to reform and migrate over the pipeline resulting in a variation in pipeline cover with time (Langhorne, 1980). In areas of mobile sand self-burial can provide a cheap and effective way of protecting and stabilising a pipeline, this can be enhanced with the use of a spoiler placed along the top of the pipe (see Section 7.4.9).

Protection from undermining by scour can be provided with a filter layer overlain by rock to the midheight of the pipe (Hjorth, 1975), more complete protection being afforded by full coverage.

Grace (1980) reviews the use of native material or rock to backfill trenched pipes. Steel slag has proved an effective protection and scour hole stabilisation material for pipelines (Dahlberg, 1981).

Water jetting is commonly used as a technique for sea bed excavation and trenching. Water jets installed on the underside of jack-up caissons have been used to achieve the required penetration of caisson foundations in sandy sediments (Song *et al.*, 1979) and this approach, which fluidises the sea bed sediment temporarily reducing its bearing strength, can be adapted to increase the embedment of other structures into the sea bed.

An analysis of the role of object embedment in the stability of sea bed debris is presented in the report prepared by Wimpey Offshore (1990).

6.5. SANDBAGS

Sandbags are routinely used as a stabilisation method for scour holes but require significant diver time to install them and are only really effective for a relatively short period of time as they can quickly become undermined (Watson, 1979).

Sandbags are used as an effective short-term stabilisation measure for jack-up rig footings (Lyons and Willson, 1986; Sweeney *et al.*, 1988). Sand bags or grout filled bags are also used to underpin pipeline freespans, but may then require a protective covering of gravel or mattresses to provide protection to the underpinning material and to stabilise the bed around the pipeline (Van Dijk, 1980), prior to the installation of a protective layer of rock or mattresses.

The use of clusters of sand bags held in nylon bags has been trialled at the 'Nordsee' platform (Maidl and Schiller, 1979) but the resulting scour around the edges of the cluster led to their displacement away from the area which they were installed to protect. Failure of some of the sandbag clusters allowed scour trenches of up to 0·8 m depth and 3 m length to occur around the periphery of the structure. The service life of the sandbag clusters was estimated as 1 to 2 years.

6.6. FLOW ENERGY DISSIPATION DEVICES

Flow energy dissipation is one of the most attractive methods for reducing the scour potential as it tackles the source of the scour itself. It is commonly used on dam or barrier overflow spillways. Methods of reducing the scour at piled structures include the installation of energy dissipating devices around the pile or pile cluster: the reduced flow energy around the base of the pile reduces the potential for sediment transport (scour) and enhances the local deposition of sediment.

The use of small piles placed in front of a larger pile to break up the flow has been shown to produce a small reduction in scour at the pile (see Breusers *et al.*, 1977). Some success has been had in reducing the scour at breakwaters through the installation of a field of wave energy dissipating blocks placed on the sea bed in front of a breakwater (e.g. Funakoshi, 1994). The local interaction of these blocks with the sea bed and their long-term stability should be considered (Nago and Maeno, 1995).

The same kind of benefits can be gained from the installation of mats studded with neutrally buoyant fibres, i.e. artificial seaweed (e.g. Hindmarsh, 1980; Maidl and Stein, 1981). These fronds are installed so that either the fronds simulate natural seaweed, i.e. fixed at the bed, or they hang down from a support frame. The speed of the water flow through the seaweed canopy is reduced (by up to 80%, Hindmarsh, 1980) and sediment suspended in the water column can settle to the bed where continued deposition can eventually form a bank.

The major operational difficulties with artificial seaweed mats are to ensure that the fibres become fully open, otherwise they have minimal effect, and to ensure that the foundation anchors and strops have adequate strength. The fibres are only just neutrally buoyant and fouling can reduce their effectiveness. Whilst this material is successful for erosion control by currents it is less successful at controlling wave induced scouring and shoreline erosion. Experiences with artificial seaweed in the US and Europe are described by Rogers (1987).

Watson (1979) reported on the success of a Sedimentary Deposition Device (SDD) installed around the legs of jackets and around risers in the southern North Sea to cover existing scour holes. This consisted of a 10 m diameter heavy polypropylene sheet perforated at the perimeter and guyed and anchored to the sea bed. The SDD prevented the horseshoe vortex from eroding the bed around the jacket legs and the reduction in flow energy within the SDD caused sediment to fall out of suspension, which led to a local build-up in the level of the bed in the scour hole. A similar device used Antiscour Device nets, comprising a fine-mesh nylon net. The mode of operation was similar to the SDD but it was found to be less durable and robust than the SDD.

6.7. SOIL IMPROVEMENT

Various mechanical and chemical methods for achieving the artificial stabilisation of soils in port and harbour engineering (in Japan) prior to the construction of bottom founded structures have been reviewed (CDIT, 1986). These techniques may have some beneficial effects in the reduction of scour by enhancing the strength of the soil and hence its resistance to erosion.

Experience has been gained in the Netherlands on the installation of heavy concrete piers in tidal waters for the storm-surge barriers constructed along the sandy North Sea coast, e.g. in the Eastern Scheldt where the fine sand bed was improved and then protected with mattresses of various kinds to prevent scouring and loss of foundation material. The scour problem can be removed altogether by excavating the soil to bedrock or an inerodible stratum (and backfilling with a scour resistant material if necessary).

Scour case studies

7. Scour case studies

7.1. AIMS

This Chapter serves two main functions; the first is to review and synthesise some of the results of published scour investigations and the second is to summarise (where available) the most suitable prediction methods for calculating scour of the sea bed sediment locally to marine structures. The results from laboratory tests, numerical–analytical modelling and field measurements obtained worldwide have been covered to indicate the level of knowledge on scouring. To simplify things, the text is divided by structure type and subdivided again where possible to deal with scour development in current only, wave only and wave–current flows. Most of the studies and predictive approaches refer to the scour that develops on an unprotected bed due to the action of the flow. Scour is treated primarily as an extension of sediment transport theory rather than from a geotechnical point of view.

The coverage of this Chapter is as follows:

- single vertical pile
- multiple pile groups
- horizontal pipelines
- large volume structures
- sea walls
- breakwaters
- free settling objects
- jack-up platforms (spud cans)
- miscellaneous (jet scour, vessel induced scour, wrecks).

7.2. SINGLE VERTICAL PILE

The scour around a slender cylindrical pile diameter D on a sand bed in a steady unidirectional flow with velocity U and flow depth h has been examined by many authors in relation to river engineering (Breusers and Raudkivi, 1991). The scour development around this simplest of structures provides a benchmark that can be adopted and extended to more complex cases with different shapes, time-varying flow, etc. The pile is considered to be slender when the pile diameter to water depth ratio $D/h < 0{\cdot}5$, practically this is the case for many offshore jacket structures, outfall and intake risers, and piled structures used in the coastal region.

7.2.1. *Scour pattern in steady flow*

Laboratory observations indicate that initially the scouring of sediment takes place in two shallow depressions situated at 45° either side of the centreline of thc cylinder. These observations correlate closely to the pattern of shear stress amplification measured by Hjorth (1975) (Figure 7). With time the scoured areas coalesce to form a conical shaped hole around the periphery of the entire cylinder of nearly constant depth and with side slopes at near the angle of repose for the sediment bed material (generally 25° to 35° for sand) (see Figure 22—flume drained for photograph). Scour depressions which shallow in the streamwise direction are formed either side of the cylinder (Figure 5). The sediment removed from around the pile is deposited between these two depressions in the 'shadow' of the pile and at the downstream extent of the wake vortex system (see Section 2.3).

Based upon the investigations in Chapter 2 it appears that it is the magnitude and extent of the flow field that controls the depth of scouring S in the region ±60° from the flow axis around the periphery of the cylinder. The final extent x_s of the scour pit from the cylinder wall is controlled by the angle of repose of the sand ϕ_i, i.e. $x_s = S_e/\tan\phi_i$. Typically (Figure 23) $S_e = 1{\cdot}3D$ and $\phi_i = 30°$ giving

$$x_s = \frac{1{\cdot}3D}{\tan 30} \rightarrow 2{\cdot}25D$$

Thus the overall diameter of the scour pit will be around $6D$.

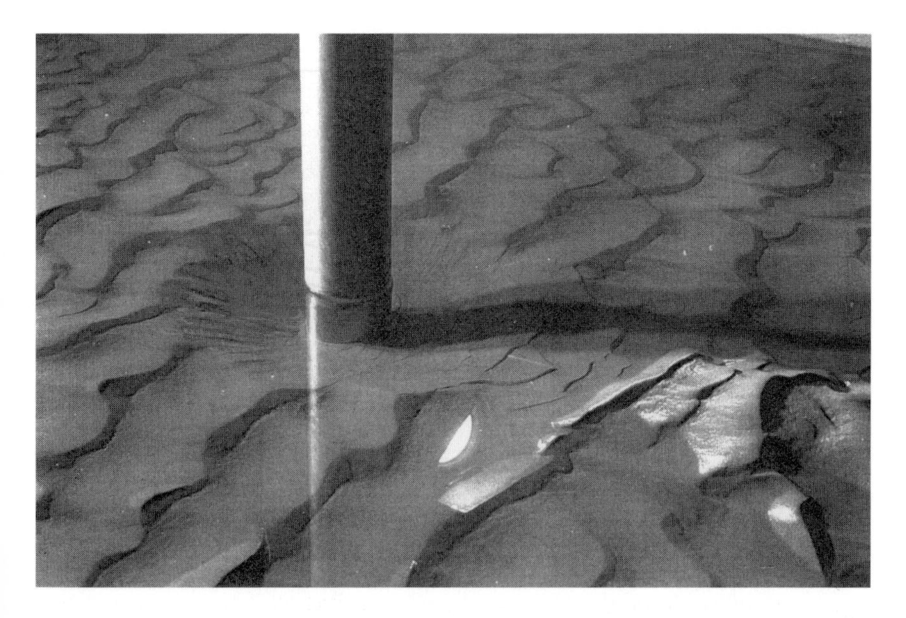

Figure 22. Scour around a pile due to steady flow from left to right: clear-water scour, $U/U_{cr} \approx 0{\cdot}9$ (reproduced by permission of BP International Ltd)

7.2.2. Scour depth in a steady flow

The depth of scour at the pile is generally assumed to scale with the diameter of the pile D (Carstens, 1966; Breusers, 1972). The equilibrium value of the scour depth increases linearly with increasing shear stress from zero to a maximum value between the points at which τ_0 is equal to τ_{cr}/M until $\tau_0 = \tau_{cr}$ (Figure 23). For larger values of τ_0 the variation in the equilibrium scour depth with increasing τ_0 is more or less constant, possibly with a 10% reduction (Breusers and Raudkivi, 1991) or a periodic fluctuation related to the passage of ripples through the scour hole. The fluctuation in the scour depth due to the passage of large bedforms should be taken into account as the scour depth could fluctuate by ±50% of the bedform height. The bedform height can be calculated using the methods given in Soulsby (1997).

The indications from published results (Clark *et al.*, 1982) are that the maximum equilibrium scour depth varies from less than $S_e = 1{\cdot}0D$ up to $S_e = 2{\cdot}3D$ for different experiments. Breusers *et*

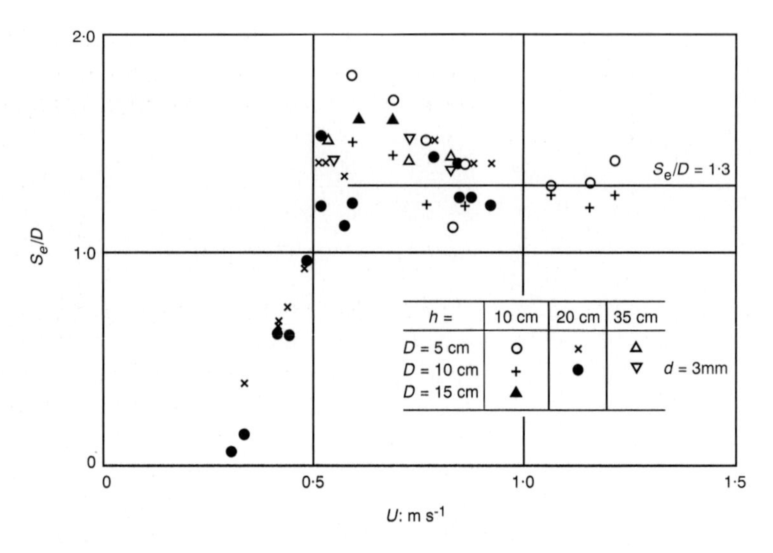

Figure 23. Variation of scour depth with flow speed (reproduced from Breusers, 1972, by permission of Delft Hydraulics)

al. (1977) obtained a maximum value of $S_e/D = 1{\cdot}5$ from their compilation of data but recommended taking the design value of 2·0 to be 'on the safe side' in Equation (35). Ettema (1990) has recommended that a conservative scour depth of $S_e = 2{\cdot}4D$ should be taken for design purposes.

The design equation of Breusers *et al.* (1977) is often used to predict the equilibrium scour depth because it contains coefficients for the ratio of the ambient to threshold flow speeds U/U_{cr}, the water depth h to pier diameter D ratio and the pile shape for rectangular piles:

$$S_e/D = c_1 c_2 c_3 c_4 \qquad (35)$$

where: $c_1 = 0$ if $U/U_{cr} < 0{\cdot}5$; $c_1 = (2{\cdot}0U/U_{cr} - 1)$ for U/U_{cr} between 0·5 and 1·0; $c_1 = 1$ for $U/U_{cr} > 1{\cdot}0$; $c_2 = 2{\cdot}0 \tanh(h/D)$, with 2·0 being the *design* value of the constant proposed by Breusers *et al.*

The other coefficients c_3 (order 1) and c_4 (order 1) are related to the effect of pile shape (discussed below) and the length to breadth diameter of rectangular piers (Hoffmans and Verheij, 1997). In addition, some correction will need to be made for the effect of alignment of the pier to the flow (Hoffmans and Verheij, 1997).

When individual sets of measurements are examined they tend to produce quite reasonable trends for the variation of S_e/D with increasing velocity (10% error) but when a compilation of data is examined the scatter is much larger. Imberger *et al.* (1982) collected data from laboratory tests for two different sediment and pile diameter combinations and combined their results with the data from five previous studies (Figure 24). The resulting graph suggested a scatter of 50% from a best fit line to the scour depth data as a function of the ratio u_*/u_{*cr}, where the shear velocity u_* is defined as $(\tau_0/\rho)^{0\cdot5}$. This scatter is most likely due to the uncertainty in the values of u_* and u_{*cr}.

For live bed scour the equilbrium scour depth is taken from Figure 23 to be

$$\frac{S_e}{D} = 1\cdot3, \text{ when } \theta \geq \theta_{cr} \text{ or } U_c \geq U_{cr} \tag{36}$$

and the expression for c_1 can be used when $\theta < \theta_{cr}$ rewritten in terms of the bed shear stress. Thus for clear water scour the equilibrium scour depth is given by

$$\frac{S_e}{D} = 1\cdot3\left[2\sqrt{\frac{\theta}{\theta_{cr}}} - 1\cdot0\right], \text{ when } 0\cdot25 \leq \frac{\theta}{\theta_{cr}} < 1 \text{ or } 0\cdot5 \leq \frac{U_c}{U_{cr}} < 1 \tag{37}$$

and when $\theta < \theta_{cr}/M$ (with $M = 4$ for a single pile) there is no scour. Equations (36) and (37) are depicted in Figures 3*a* and 24.

7.2.3. *Time variation of scour*

For a given set of environmental conditions the scouring of the sandy sediment at structures initially occurs rapidly but then approaches its ultimate (equilibrium) value asymptotically (Figure 3*b*). The rate at which the scouring of sediment takes place from around the pile is related to the divergence in the sediment transport rate for the control volume around the structure which results from the disturbance to the flow field.

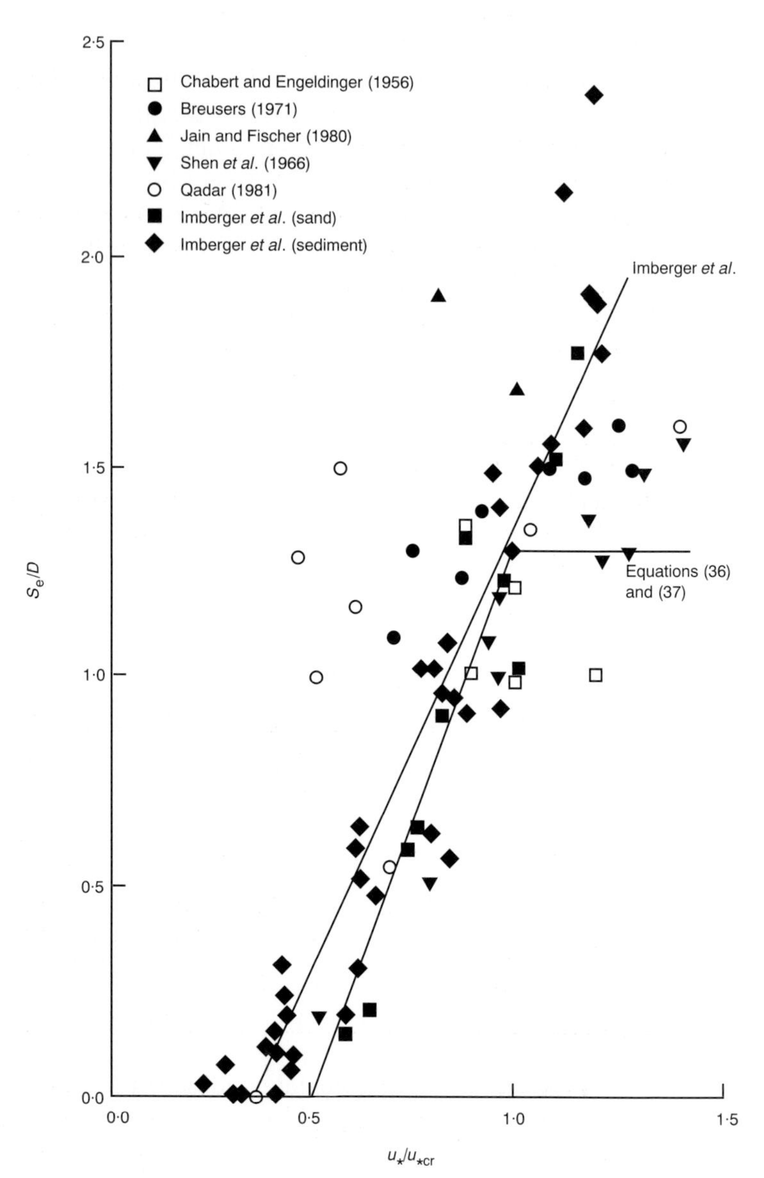

Figure 24. Variation of scour depth with u_*/u_{*cr} and the line of best fit added by the authors plus Equations (36) and 37 (reproduced from Imberger et al., 1982, by permission of the ASCE)

Measurements of the increase in depth S of the scour pit with time t around a fixed vertical pile in a steady current and waves (Sumer *et al.*, 1992a) have been found to fit well to Equation (3)

$$S(t) = S_e\left[1 - \exp\left(-\frac{t}{T}\right)^p\right]$$

with $p = 1{\cdot}0$ for both wave and current induced scouring.

The dimensionless time-scale of T^* for $\theta > \theta_{cr}$ can then be predicted using the Shields parameter for the ambient flow (Equation (5b)) via Equation (5a).

$$T^* = A\theta^B$$

where the constants A and B have been determined from the data of Sumer *et al.* (1992a), see Figure 25, as 0·014 and −1·29 respectively. For a particular sediment characteristic and pile diameter the value of T required in Equation (3) can be obtained via Equation (4) and hence the time development can be predicted.

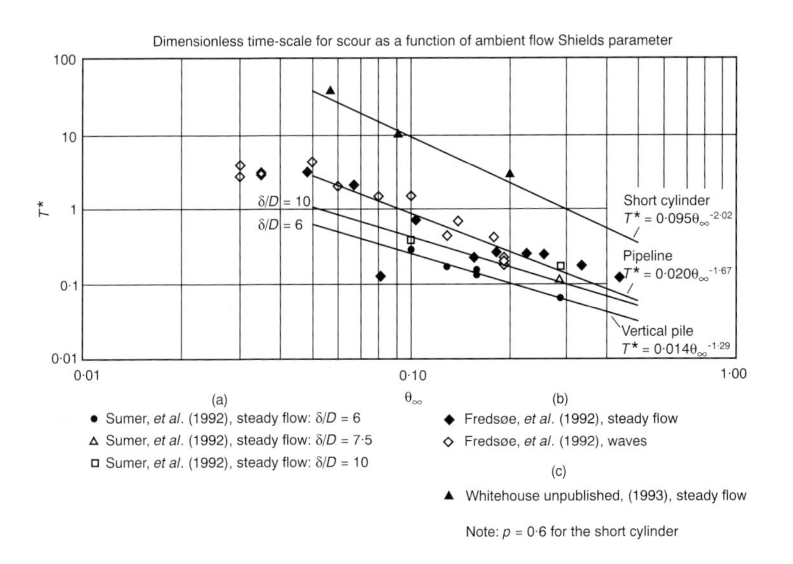

Figure 25. Time-scale for scour ((a) vertical pile, (b) pipeline and (c) short horizontal cylinder; sources as indicated)

Figure 25 shows the vertical pile results for three ratios of the boundary layer thickness δ (actually the water depth h in Sumer *et al.*'s experiments) to the pile diameter. The time-scaling constants used above were determined for the data at $\delta/D = 6$ as this should correspond to the case of 'deep' water (see Section 7.2.4). Based upon the limited amount of data available for deeper water conditions, $\delta/D = 7{\cdot}5$ and 10, the constant A could be about 50% larger.

7.2.4. Influence of water depth

Almost certainly some of the scatter in published scour depth plots such as Figure 24 will be attributable to the effect of shallow water although Imberger *et al.* (1982), for example, claim to have corrected for this in their plot (Figure 24). The effect of water depth on the scour depth at cylindrical piles is generally agreed to be negligible for values of the ratio $h/D > 3$ (see Breusers, 1972; May and Willoughby, 1990; Yanmaz and Altinbilek, 1991), i.e. tanh $h/d \approx 1$ when h/D equals 3 (Equation (35)). Based upon the shear stress amplification data discussed in Section 2.5.1, deep water would correspond to $h/D \geq 5$. Although the measured scour depth is reduced for values of $h/D < 3$ this reduction factor is not of practical significance for most piled structures in the marine environment as typically $h \gg D$.

7.2.5. Pile shape

The effect of pile cross-sectional shape on the equilibrium scour depth has been studied in steady flow experiments (Breusers and Raudkivi, 1991; May and Willoughby, 1990). Taking the scour depth at a circular pile as reference the equilibrium scour depth multiplier for square piles has been determined as 1·3, e.g. S_e (square) = $1{\cdot}3S_e$ (circular). Attempts at streamlining, for example as for bridge piers in rivers, can achieve a reduction in the scour depth (multiplier = 0·75) but this reduction cannot generally be taken advantage of in the offshore environment due to the effect of the rotary tidal current. The effect of cross-section shape on wave-induced scour is discussed in Section 7.2.9.

7.2.6. *Sediment gradation*

As a direct result of the need to make scour predictions in rivers where the bed has become armoured by a layer of coarser material, a large amount of work has been performed in New Zealand to determine the effect of sediment size and gradation on the scour depth (Breusers and Raudkivi, 1991). The scour depth will not be limited by the grain size providing $D > 50d_{50}$. For clear-water scour the maximum value of the equilibrium scour depth (at $\theta \approx \theta_{cr}$) is inversely proportional to the geometric standard deviation of the particle size distribution, σ_g, and also depends to some extent on whether the sediment is ripple forming, i.e. $d_{50} < 0{\cdot}8$ mm. Breusers and Raudkivi suggest that any specific value of S_e/D for clear-water scour is unique to a given value of σ_g. In the sea, where the grain size of the natural bed material will be very much smaller than the typical diameter of the pile and the bed material tends to be well sorted, the limiting influence of grain size and grading on scour is probably negligible.

7.2.7. *Cohesive sediment*

The individual sand particles will be subjected to cohesive forces when the sediment contains more than 10% by weight of cohesive silt sized material. The effects of the cohesion are to increase the resistance of the bed to scour (Mitchener *et al.*, 1996; see Appendix 2) and to produce a more complex shaped scour hole with sideslopes that are steeper than the angle of repose for the constituent sand. The mode of sediment transport for layered mud:sand beds will alternate between suspended and bedload transport as different layers are eroded.

7.2.8. *Scour pattern in wave flow*

Where the cylinder diameter is less than 20% of the local maximum wavelength of the sea surface waves (or bottom orbital velocity amplitude) the effects of wave diffraction are minimal and the scour pattern proceeds firstly with the development of scour depressions either side of the pile, at 90° to the centreline of the cylinder (Sumer *et al.*, 1992b; Abou-Seida, 1963).

Ultimately the scour pattern is not too dissimilar to the current only case, but without the pair of shallow streamwise depressions extending downstream. The wave-induced scour pattern for a square pile is similar to the circular pile case (Sumer *et al.*, 1993).

7.2.9. *Scour depth and development in wave flow*

The scour depth in oscillatory wave flow has been examined in several papers but the most comprehensive study was by Sumer *et al.* (1992b). Most authors agree that the local scour around a pile due to waves is smaller than the steady current value. Based upon extensive laboratory data, Sumer *et al.* correlated the equilibrium scour depth with the Shields parameter and the Keulegan Carpenter (KC) number (Equation (11)). Their live-bed scour data ($\tau_{w} > \tau_{cr}$) demonstrated firstly that the equilibrium scour depth due to waves is considerably smaller than due to a steady current, i.e. $S_e/D < 1$ for $KC < 55$, and secondly that there is a good correlation between the scour depth and the KC number (Figure 26). The relationship was approximated by the equation:

$$\frac{S_e}{D} = 1{\cdot}3\{1 - \exp[-0{\cdot}03(KC - 6)]\}, \text{ for KC} \geq 6 \qquad (38)$$

The scour depth is negligible for $KC < 6$ due to the non-existence of the horseshoe vortex and shed vortices. The scour depth for large KC numbers, greater than 100, is similar to the steady flow value of $1{\cdot}3D$. This limiting value of KC below which the influence of the reversing flow becomes significant was also quoted by Breusers and Raudkivi (1991).

Tests performed by Sumer *et al.* with $KC > 1000$ indicate that the maximum scour depth in a tidally reversing flow is also similar to the steady flow value (Equation (36)), although the role of backfilling of scour holes on alternate tides is not explicitly addressed.

The effect of pile cross-sectional shape has been studied by Sumer *et al.* (1993) in tests on circular piles and square piles at orientations of 90° and 45° (corner into flow). The data indicate that the equilibrium scour depth for $KC > 100$ approaches a constant value of $S_e/D = 1{\cdot}3$ for circular piles and $S_e/D = 2$ for square piles, i.e. a 50% increase in the wave-induced scour depth

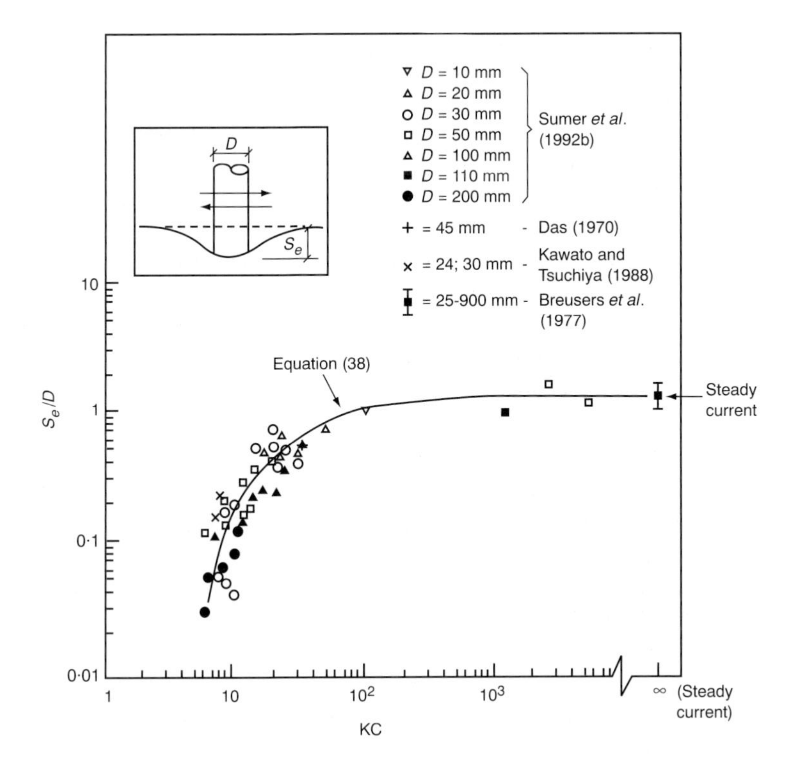

Figure 26. Correlation between equilibrium scour depth at circular pile and Keulegan Carpenter number: live-bed conditions (reproduced from Sumer et al., 1992b, by permission of the ASCE)

at a square pile as opposed to 30% for the steady flow case (Section 7.2.5). The variation in scour depth with KC number for the different pile shapes/orientations was more complicated for KC < 100.

The time-scale with $p = 1{\cdot}0$ in Equation (3) for scour development at cylindrical piles has also been shown to depend upon the KC number as well as the Shields parameter. Sumer *et al.* (1992a) proposed the following expression for live-bed scour:

$$T^* = 10^{-6}\left(\frac{\mathrm{KC}}{\theta}\right)^3 \tag{39}$$

where θ is derived using the maximum amplitude of the bottom shear stress τ_{w}.

7.2.10. Scour depth in wave–current flow

The effect on the local scour depth obtained by adding waves onto a steady current is still unclear since some investigators have found that the waves reduce the scour depth (Abou-Seida, 1963; Bijker and de Bruyn, 1988) and others that the scour depth is not increased above the current-only value (Clark and Novak, 1984). It is generally agreed that the rate of scour in wave–current flow is enhanced (Chow and Herbich, 1978; Machemel and Abad, 1975; Clark and Novak, 1984). Kroezen *et al.* (1982) state that the influence of waves on the scouring rate is of great importance when the steady current alone is too weak to cause scour.

Bijker and de Bruyn (1988) found that the wave plus current shear stress was increased more in the area upstream of the pile than in the vicinity of the pile and that the enhanced transport of sediment towards the pile led to a reduction in the scour depth. Bijker and de Bruyn (1988, Series I and II tests) found for the live-bed case the maximum scour depth to pile diameter ratio S_e/D was 1·3 for a current alone and 1 to 1·1 for a current and non-breaking waves. Chow and Herbich (1978) found the scour depth in collinear wave–current flow was 10% deeper than the current-only case. Clark and Novak (1984) determined the maximum equilibrium scour depth (live-bed scour) in laboratory flume experiments with collinear waves and currents as $1{\cdot}7D$. Armbrust (1982) found that for large (relative to the bottom orbital velocity) current speeds the wave action appears to increase the scour depth substantially. Thus it appears that the ratio of wave to current speed as well as their magnitudes influences the scour development in wave–current flow.

7.2.11. Scour pattern in wave–current flow

The local scour pattern formed by a collinear wave–current flow is probably similar to the current-alone case, although the overall scour pattern produced by wave–current flows crossing at other angles will be more complex. This is because the net transport of sediment entrained from the bed by the wave motion will be more or less in the direction of the tidal current. Posey and Sybert (1961) were able to reproduce the dishpan scour phenomenon but these tests were completed with large (standing) waves and currents crossing at right angles and a fine, lightweight bed material.

Bijker and de Bruyn (1988) found the scour hole extent x_s was $3D$ upstream and $5D$ downstream for the current-alone case and this was increased by wave–current flow to $4D$ and $6D$ respectively.

7.2.12. *Breaking waves*

The main effect of breaking waves in the sea is to provide an additional source of turbulence in the water column, if the water depth is shallow enough to cause wave breaking (Southgate, 1995), and the extra turbulence will enhance the suspended sediment-carrying capacity of the flow. Bijker and de Bruyn (1988) measured greater scour depths (up to $S_e/D = 1{\cdot}9$) when breaking waves were superimposed on a current, i.e. 46% larger than the current-only situation. The available data on this topic are sparse.

7.2.13. *Storm effects*

In general there is a lack of published investigations on the effect of storms on the scour at piles. Interesting experimental results have been reported by Di Natale (1991) who measured the time variation in the scour depth at a pile during the passage of a storm. The results from the experiments showed that the peak of the scour development was out of phase with the peak of the storm and that the maximum scour depth recorded during the storm was considerably different (4 to 7 times greater) than was observable at the end of the storm, i.e. backfilling had taken place after the storm had passed.

7.2.14. *Effect of resistant bed layer*

Where a pile of diameter D has been driven in an area with a veneer of erodible sand that is less than $1{\cdot}0D$ to $2{\cdot}4D$ thick (Section 7.2.2) overlying a resistant layer, e.g. stiff clay, the depth of scouring can be limited. If the resistant layer outcrops upstream of the installation the sediment load in the flow as it approaches the pile might be lower than the equilibrium carrying capacity of the flow. The significance of this situation is that the flow will have a larger scouring potential than if it was already carrying sediment.

7.2.15. *Influence of scour*

The reduction in pile fixity due to the scour around single piles is not as large as the reduction due to the scour resulting from group effects, i.e. global scour (Diamantidis and Arnesen, 1986).

7.2.16. *Field observations*

Based upon published empirical data and field experience from the Netherlands, de Wolf *et al.* (1994) predicted scour depths of between $2D$ and $3{\cdot}1D$ for the single pile of the Westhinder research platform 32 km off the Belgian coast.

7.3. MULTIPLE PILE GROUPS

In the case of pile clusters, group effects become important as does the angle of orientation of the pile cluster to the prevailing direction of waves and currents. The main factors to consider are flow interference leading to enhanced flow speeds or turbulence at adjacent piles in some cases and sheltering of piles in others.

7.3.1. *Linear arrays of piles*

The effect of pile spacing on the scour depth in a steady current has been investigated in the laboratory. From experiments Breusers (1972) and Hirai and Kurata (1982) found that the scour depths at individual cylindrical piles in a cross-flow array were the same as for the single pile provided that the pile spacing (distance between pile centres) was greater than or equal to $6D$. Basak *et al.* (1975) (referenced in Breusers *et al.* (1977)) determined the critical spacing for square piles as $5D$. Experimental data (Hirai and Kurata, 1982) showed that the scour depth increased as the spacing of two piles perpendicular to the flow direction was reduced from $6D$ to $2D$). For a pile spacing of $2D$ the scour depths at the sides of the cylinders were increased by approximately 40% from the single pile value. As the pile gap decreased further the relative scour depth increased until it exceeded two times the value for the single pile (i.e. proportional to the projected area of the pair of piles).

Interaction effects have also been observed for piles aligned parallel to the flow. Experiments have shown that for pile spacings in the range $1{\cdot}5D$ to $4D$ the scour depth around the upstream pile was increased by 10% to 20% (Hirai and Kurata, 1982). However, when the spacing was increased to $6D$ the scour depth at the aft cylinder was reduced to 60% of the single cyclinder value as a result of the deposition of scoured sediment taking place in this region. Breusers and Raudkivi (1991) report the data of Hannah (1978) for scour in a steady current behind (1) a single pile, (2) a group of piles and (3) two piles, both in line and at an angle. The results for the two piles in line with the flow are similar to Hirai and Kurata except the large deposition at the aft pile was not observed.

The results from experiments with squat piles reported by Hirai and Kurata (1982) and Vitall *et al.* (1994) showed that there is a significant (50%) reduction in the scour depth as the pile height is reduced to only 10% of the water depth.

7.3.2. Pile clusters

The scour interaction of groups of piles can be estimated in many cases from the published literature. The scour around, for example, the pile cluster legs of braced jacket structures with mudmats should be investigated on a case by case basis.

Experiments on the scouring pattern obtained both within and downstream from 3×3, 5×3 and 7×3 groups of square, octagonal and hexagonal cross-section piles have been reported by Mann (1991). These experiments used a thin layer of sand on the bed of a flume as a tracer material to indicate the area of the bed that was scoured from around the pile groups and the results were presented in terms of a scour ratio (SR = ratio of scoured area of bed to group area of pile array). The observed value of SR for the 3×3 pile group in head-on flow varied with the pile spacing (centre to centre) as follows: (spacing:SR) = ($3D$: 1·8),

($6D$: 0·9) and ($9D$: 0·3). The variability in the scour pattern under different angles of attack was also investigated. The value of SR tended to be larger for the case where the array was rotated 45° to the approach flow.

The clear-water scour depth at a group of 3 cylindrical piles angularly spaced at 120° was studied by Vittal *et al.* (1994) with the gap set so that any one of them could just pass through the gap between the other two (i.e. gap = $2D$). The scour depth at the pile group due to a steady current was found to be about 40% smaller than the scour depth at a single pile with a diameter equal to the diameter of a circle circumscribing the pile group. The scour depth varied by only 6% with variations in the angle of attack of the flow.

The scour at groups of 3, 4 and 6 piles was investigated in a wave flume by Chow and Herbich (1978) (also see Herbich *et al.*, 1984) and the results obtained indicate that there was an increase in the scour depth as the pile spacing was reduced from $8D$ to $4D$. Data for the time-variation of scour were also presented. The results from these experiments are presented as scour depth non-dimensionalised with the water depth and would need to be re-worked with scour depth related to pile diameter to present them in a consistent format.

The techniques for predicting scour under fixed pipelines discussed in Section 7.4 can be applied to predicting local scour under the lowermost horizontal framing member diameter D of jackets, with an initial gap e_0 in the range $0 \leq e_0/D \leq 2$ (Sumer and Fredsøe, 1990).

7.3.3. *Field observations*

The potential for large overall scour depths around piled jetties is illustrated well by the bathymetric data for the research pier at Duck, North Carolina illustrated in Figure 2. Field observations of scour at piled structures in the coastal zone can be made at low tide although the influence of shallow water effects in modifying the scour holes local to the pile could be significant as the tide recedes prior to observation.

The presence of lattice work and cross-bracing in the jacket structures used offshore leads to an increase in the levels of flow turbulence around the platform. Watson (1979) reported that the lowermost brace of some jackets lay on the sea bed and the associated increase in turbulence levels had been postulated as a possible contributory factor in the generation of dishpan scour shown schematically in Figure 1 (Posey, 1970; Tesaker, 1980). The scour development at jacket structures in the Gulf of Mexico is documented by Posey and Sybert (1961), with the scour depth measured as the distance from the lowermost bracing level to the bottom of the scour pit. Average scour depths beneath five different structures varied from 0·6 to 3·6 m.

7.4. PIPELINES

The consequences of scouring of sediment from around offshore pipelines has received a large amount of attention as the total length of subsea pipelines in the North Sea and elsewhere has continually been extended and the pipes installed in the early days of oil exploration are ageing. It is now estimated that the combined distance of pipelines in the North Sea associated with hydrocarbon exploration alone is greater than 15 000 km (Kyriacou, A., personal communication, 1996). Pipelines are used in coastal situations for outfalls and intakes and the landfall of offshore pipelines. They are either buried on installation or fixed at some height relative to the local bed level at the time of installation.

Pipelines installed on the sea bed are susceptible to damage arising from environmental forces (waves and currents loading, flow induced vibration, sediment loading, e.g. Sumer *et al.*, 1988a and b) or the influence of man's activities (e.g. trawl board or anchor loading; CIRIA/CUR, 1991; Verley *et al.* 1994). The presence of the pipeline on the sea bed disturbs the local flow field producing an imbalance in the local sediment budget which leads to scouring of the sea bed and the generation of free-spans. The risk of damage from external forces or the self-weight sagging of the pipe is increased where lengths of pipeline are free-spanning between mounds of unscoured sediment. Therefore, the scouring needs to be predicted and monitored to minimise the risk of damage.

In some cases the scour at pipelines can be beneficial where this leads to the burial of the pipeline into the sea bed. An extensive programme of research on the self-burial of pipelines has been carried out in The Netherlands because pipes located in the highly mobile sandy area of the southern North Sea are susceptible to self-burial (Figure 27). Self-burial reduces the costs associated with installing a pipeline, because it can minimise or entirely remove the need to trench and backfill a pipeline during installation. The potential saving on installation costs arising from the trenching and backfilling of pipelines in

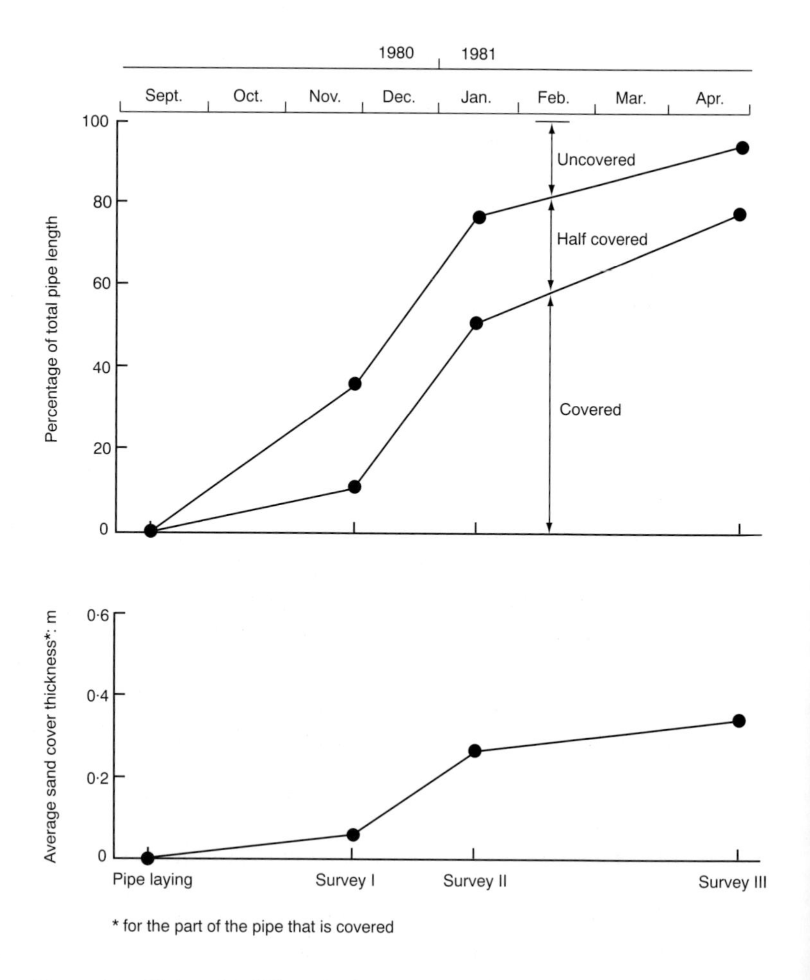

Figure 27. Observed self-burial of 12 inch pipeline in the southern North Sea (reproduced from Kroezen et al., 1982, by permission of Delft Hydraulics)

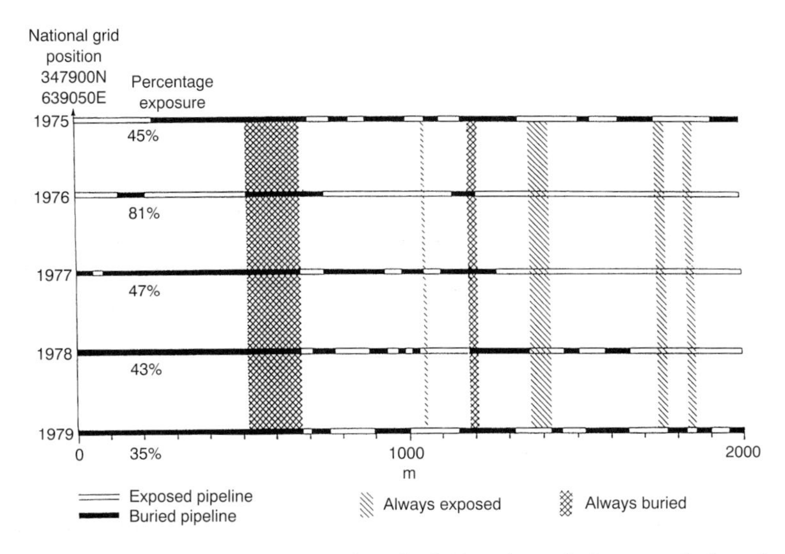

Figure 28. Temporal variation in burial of 30 inch trunk line routed through Haisborough Sand sandwave field (reproduced from Langhorne, 1980, by permission of the Institute of Oceanographic Sciences/Information Services, Southhampton Oceanography Centre)

the mid-1980s was of the order of £2 million for a 25 km pipeline (ICE, 1985).

Most pipeline investigations are based upon a 2-dimensional slice taken through an infinitely long pipeline but it must be borne in mind that in reality the pipe scour problem is a 3-dimensional problem (Sumer and Fredsøe, 1993). The pipeline will undulate over the sea bed, and will invariably not run in a straight line, and the gap beneath the pipe and the percentage of the pipe casing covered by sand will vary along its length (as illustrated in Figure 28).

7.4.1. Field experience

Kroezen *et al.* (1982) reported observations of a 12 inch pipeline placed in 30 m of water on a sandy substrate in the southern North Sea. Within 4 months of laying the interaction of the pipeline with the flow had scoured a trench $3D$ deep and $50D$

wide and 50% of the total length of the pipeline had been totally covered by sand to a depth of 0·3 m. After 7 months the pipeline had become totally buried over 70% of its length (Figure 27). Subsequently, the results of this study were combined with data from small and large scale laboratory tests and supplemented with investigations as to how spoilers fitted to the pipe can enhance the burial potential.

Reports from regions with bedforms have indicated significant annual variations in the exposure of the pipeline due to the migration of sandwaves. Langhorne (1980) reported a comprehensive study of the pipeline–sandwave interaction at the northern end of the Haisborough Sand sandwave field where five gas lines pass *en route* from the southern North Sea gas fields to the shore terminal at Bacton, Norfolk, UK. Detailed summer surveys of a 2 km length of Shell's 30 inch pipeline between 1975 and 1979 (Figure 28) revealed the following information for the variation in the percentage exposure during the 5 years: 45%, 81%, 47%, 43% and 35% exposure. The locations of exposed and buried sections along the length of the line did not vary in a systematic fashion and only 8% of the length of the pipeline remained constantly buried and 7% constantly exposed.

At one location the pipeline was observed to vary during the survey from being 3·4 m above the local bed level to 0·6 m buried. Previous survey data from 1967–70 showed that the sea bed level had accreted by up to 5·5 m, increasing the protection afforded to the pipeline. Information on the variation in exposure and burial during the winter months was not available.

The results of a questionnaire survey of operators' views on the nature of scour at offshore pipelines have been summarised by Harley (1992). Field observations of scour at intertidal pipelines are discussed in Section 7.4.15.

7.4.2. *Effect of water depth*

When the upstream water depth exceeds 4 times the pipe diameter the effect on the flow field around the pipe (Chiew, 1991a) and on the scour development under the pipeline is not significant. HR Wallingford (1972) examined the prototype behaviour of pipelines at an intertidal estuarine site and considered 'deep water conditions' to be approximated when the flow depth was greater than 3 times the pipe diameter. This limit was also determined in the laboratory by Kjeldsen *et al.* (1974). From a practical viewpoint water depth effects do not need to be considered for offshore pipelines but may influence the scour development under pipelines crossing the intertidal zone.

7.4.3. *Scour profile in current*

An extensive series of laboratory flume tests on the time development of the scour hole beneath pipelines resting on sand beds have been reported by Kjeldsen *et al.* (1974). This was the first systematic study of pipeline scour and the results showed that the maximum scour depth occurred at a position just downstream of the pipeline axis and that the scour hole was asymmetric in cross-section (Figure 29). Similar results have been found by Mao (1986). The 3-dimensional (plan) scour pattern has been studied by Mao (1986).

7.4.4. *Scour depth in current*

The appropriate scale parameter for the scour depth at pipelines is the external diameter of the pipe. Kjeldsen *et al.* (1974) presented an empirical formula for the equilibrium tunnel scour depth under a fixed pipeline resting initially on the bed (e.g. for the situation shown in the lower panel of Figure 29)

$$S_e = 0{\cdot}972\left(\frac{U^2}{2g}\right)^{0{\cdot}20} D^{0{\cdot}80} \tag{40}$$

This formula is valid in the range $9{\cdot}84 \times 10^3 < \mathrm{Re}_{\mathrm{pipe}} < 2{\cdot}05 \times 10^5$ ($\mathrm{Re}_{\mathrm{pipe}} = UD/\nu$). Subsequent work in the Netherlands using additional data has led to a modification of this formula to contain a moderate (negative) grain size dependence (Bijker and Leeuwenstein, 1984)

$$S_e = 0{\cdot}929\left(\frac{U^2}{2g}\right)^{0{\cdot}26} D^{0{\cdot}78} d_{50}^{-0{\cdot}04} \tag{41}$$

Both these formulae indicate that the strongest controlling factor on scour depth is the pipe diameter, and Lucassen (1984)

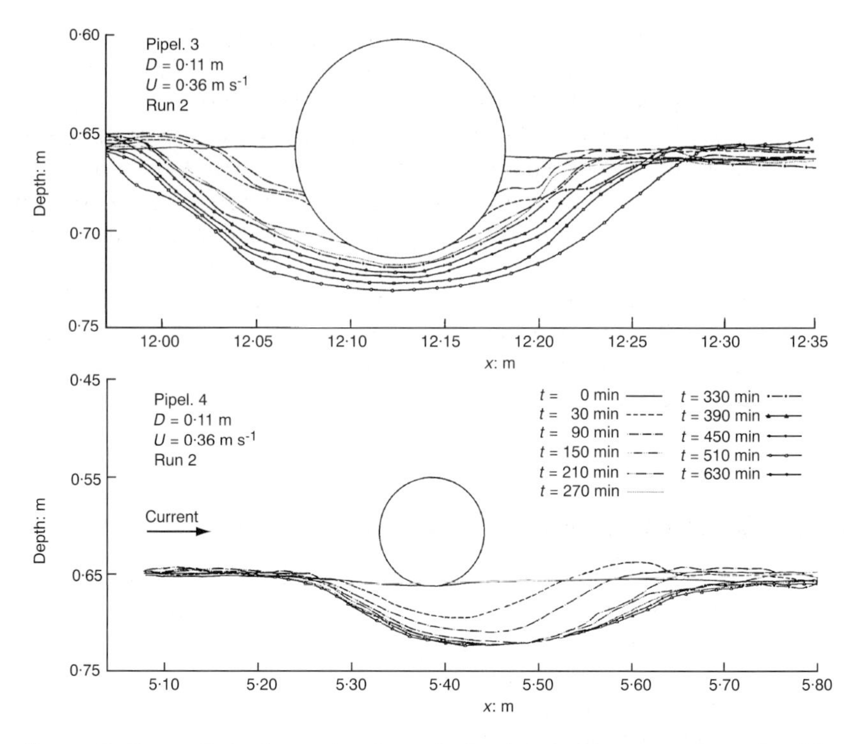

Figure 29. Scour pit development under a pipeline in a steady current (reproduced from Kjeldsen et al., 1974, by permission of SINTEF, Civil and Environmental Engineering, Trondheim)

concluded that the grain size was of minor importance in controlling the equlibrium scour depth for the live-bed case.

Sumer and Fredsøe (1990) compiled data from four previous investigations and found only a weak variation of the scour depth with pipe Reynolds number and τ_0 provided $\tau_0/\tau_{cr} > 1$ (i.e. live-bed conditions). This led to their suggestion that the average scour depth under a fixed pipeline with its lower side initially resting on the bed was approximately

$$S_e/D = 0{\cdot}6 \text{ (standard deviation} = \pm 0{\cdot}1) \qquad (42)$$

The dominance of the pipeline diameter in Equations (40) and (41) lends some support to this argument. Their compilation of data does exhibit some shear stress dependency of the ultimate scour depth for pipe Reynolds numbers larger than 10^5. As the initial gap under the pipe increases the magnitude of S_e is reduced (Mao, 1986). For a gap of $0{\cdot}25D$ the scour depth is 96% of Equation (42), 92% for $0{\cdot}3D$ and for a gap of $1D$ the scour depth is 75% of Equation (42).

Chiew (1991b) studied the scour depth in shallow flow and produced an empirical function for predicting the maximum clear water equilibrium scour depth at $\tau_0/\tau_{cr} \approx 1$. This formula contains the flow depth to pipe diameter ratio and the magnitude of the flow through the gap under the pipe. The measured scour depth was predicted with an error of $\pm 20\%$ in comparison with laboratory data.

7.4.5. *Scour in two-way flow: waves or tidal flow*

The onset of scour in waves has been studied experimentally by Sumer and Fredsøe (1991) who presented an empirical expression including KC number (Equation (11)) for the critical burial depth of the pipeline e_{cr} beyond which no further scouring occurred—e_{cr} is defined as the distance of the bottom of the pipe beneath the sediment surface.

$$\frac{e_{cr}}{D} = 0{\cdot}1 \ln \mathrm{KC} \tag{43}$$

This expression is valid for live bed scour conditions in the range $2 \leq \mathrm{KC} \leq 1000$. For pipelines with depths of burial greater than given by Equation (43) scour will not occur.

The clear-water scour depths under pipelines are only a few per cent of the pipe diameter but increase with increasing KC number when live-bed conditions are operating (Sumer and Fredsøe, 1990). From laboratory studies Lucassen (1984) also found that the scour due to waves was of minor importance compared to current scour. Whilst the live-bed equilibrium scour depth in a steady current is a weak function of the Shields parameter the scour depth in the case of waves depends strongly upon KC for the case where the pipe is initially in contact with the bed. The following equation has been proposed (Sumer and Fredsøe, 1990) based upon laboratory data and is valid in the range $2 < \mathrm{KC} < 600$ (Figure 30) for the case of live bed scour

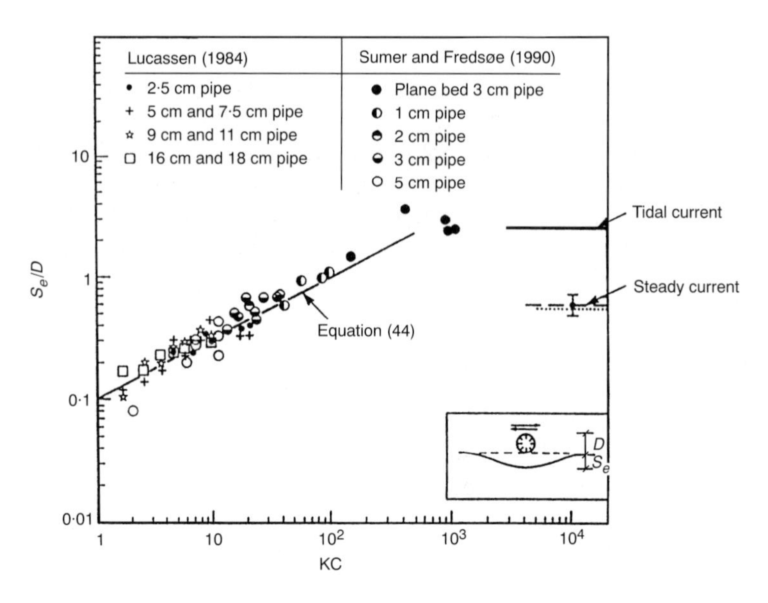

Figure 30. Equilibrium live bed scour depth under a pipeline in waves (reproduced from Sumer and Fredsøe, 1990, by permission of the ASCE)

when the bottom of the pipe is resting at the original plane bed level.

$$S_e/D = 0{\cdot}1\sqrt{\mathrm{KC}} \tag{44}$$

Sumer and Fredsøe found the scour depth was equal to the steady flow case for KC about 30, becoming deeper for increasing values of KC until $S_e/D = 2{\cdot}4$ at KC = 600. The latter value of scour depth, they suggested, was an appropriate value for the tidal flow case. Field observations (Kroezen *et al.*, 1982) have indicated that deep burial is possible in tidal flow (Section 7.4.1).

Mao (1986) and Grass and Hosseinzadeh-Dalir (1995) have examined the scour development for a fixed pipe in tidal flow (initial gap = zero) and find that, as for the situation with wave scour, the scour pit is nearly symmetrical either side of the pipe. The lee-wake scour has been observed to produce very wide (greater than 50 diameters) and deep symmetrical scour holes. Grass and Hosseinzadeh-Dalir found a good correlation between scour hole width W and depth for fixed pipes with maximum values of $W/D = 100$ corresponding to $S_e/D = 1{\cdot}8$ to 2. For $W/D = 50$ the available data indicated $S_e/D \approx 1{\cdot}6$.

It is important to note that tidally induced scour to $1D$ can occur under the pipeline when the initial gap is as large as $2D$, (Sumer and Fredsøe, 1990) increasing the total gap for a fixed pipe to $3D$. The deeper burial observed in the field is often associated with pipe sagging or vibration (see Sections 7.4.16 and 7.4.17).

7.4.6. *Scour in waves and currents*

The influence of waves and currents on scour depth has not been clearly defined: some results show reduced scour depths and others show enhanced scour. Bijker (1986) discussed the results from the Delft University of Technology. For the same bottom shear stress the scour depth under a steady unidirectional flow was always larger than that obtained under pure wave action or the combined action of waves and currents.

Based upon the limited experimental findings of Lucassen (1984) for a fixed pipe with zero initial gap it appears that the scour depth in wave plus current flow is significantly greater (order 30%) than the live bed scour due to current alone. By comparing the predictions of Equation (44), which fit well to Lucassen's wave-only live bed scour data (Sumer and Fredsøe, 1990), it is observed that the addition of a current in Lucassen's experiments (current magnitude below threshold for general bed motion) leads to an enhanced scour depth (80% greater than is predicted by Equation (44)).

Further analysis is required to produce a generally applicable formula for predicting pipeline scour due to waves and currents.

7.4.7. *Time development of scour*

From field observations Staub and Bijker (1990) reported that the tunnel erosion phase takes place very quickly after installation of the pipe on the sea bed but that the lee-wake erosion can take weeks or months to produce the ultimate scour pattern. The observations by Kroezen *et al.* (1982) indicate that a rapid scouring and burial can occur within a few months of laying the pipeline in favourable conditions.

The time development of scour is fairly well described by the negative exponential relation used for piles (Equation (3)) with $p = 1{\cdot}0$. The time-scale for scour in waves and currents alone has been investigated in the laboratory by Fredsøe *et al.* (1992) who found that the time-scale in currents as well as waves was a function of the Shields parameter θ. The larger the Shields parameter, the smaller is the time-scale. The relation expressed in terms of the non-dimensional time-scale is (Figure 25):

$$T^* = 0{\cdot}02\theta^{-5/3} \tag{45}$$

where the shear stress in θ is based upon the steady current velocity or maximum bottom orbital velocity under the wave.

7.4.8. *Pipe roughness*

The influence of the pipe coating roughness on the scour depth has been investigated in laboratory experiments by Sumer and Fredsøe (1990). The results indicated no noticeable influence on the scour depth when comparing the results from rough and smooth pipes.

7.4.9. *Spoilers*

The interest in self-burial of pipelines in sand beds has led to a number of comprehensive studies of the effect of using a spoiler fixed along the top of the pipeline to promote scour. The action of a spoiler is twofold: firstly it increases the apparent diameter of the pipeline and secondly the lee-wake turbulence levels are enhanced. Both these factors will lead to deeper and wider scour pits being generated. Van Beek and Wind (1990) have shown this to be the case in their numerical modelling of the flow field and scour with a spoiler attached to the top of the pipeline.

Laboratory studies (Hulsbergen, 1986) with a 20 inch pipe mounted with 5 inch spoilers top and bottom demonstrated that the self-burial potential of the pipeline was boosted by a factor of four with the addition of the spoilers. The first part of the test with the plain pipe held above the bed produced an equilibrium scour depth of 0·2 m (or $0{\cdot}4D$) and adding the spoilers produced a further scouring phase to a depth of greater than 0·8 m. Gökçe and Gunbak (1992) also measured an improved self-burial capacity in wave flows when a spoiler was added to the top of a plain pipe.

A detailed study of the effect of spoiler orientation on wave induced scour has been undertaken by Chiew (1993). The spoiler was effective at enhancing scour provided it lay within 30° of the vertical.

It should be noted that spoilers are only likely to be effective under the conditions where the self-burial of a plain pipeline is already expected to take place.

7.4.10. *Risers*

The scour around risers (e.g. offset thermals, J-tubes at offshore platforms) can be treated as a composite problem. The scour depth at the near-vertical riser as it leaves the sea bed can be predicted from the results of the cylindrical pile investigations (Section 7.2). If the erosion reaches down to the elbow section then the scour can propagate along the pipe away from the elbow. The scouring of sediment from under the riser can leave it unsupported and vulnerable to wave–current loading. The interaction of scour effects from the platform with the scouring around the riser might need to be considered.

7.4.11. *Cohesive sediment*

The only systematic study for the effect of cohesive silty sediment on the scour depth at pipes has been undertaken by Pluim-van der Velden and Bijker (1992), following observations of reduced scour at the Oester Grounds in the southern North Sea. They compared the results of flume experiments with sand and sand containing between 5% and 20% by weight of a natural mud to investigate the scour depth and scour pit profile in muddy sediments. The results for sand (Figure 31, upper panel) showed the equilibrium scour depth as predicted by Equation (42) and the typical scour profile.

The results of the mud:sand tests indicated that the effect of the cohesive material was to increase the value of τ_{cr}, as also reported from work on mud–sand mixtures by Mitchener *et al.* (1996). Once erosion had been initiated the low cohesion bed

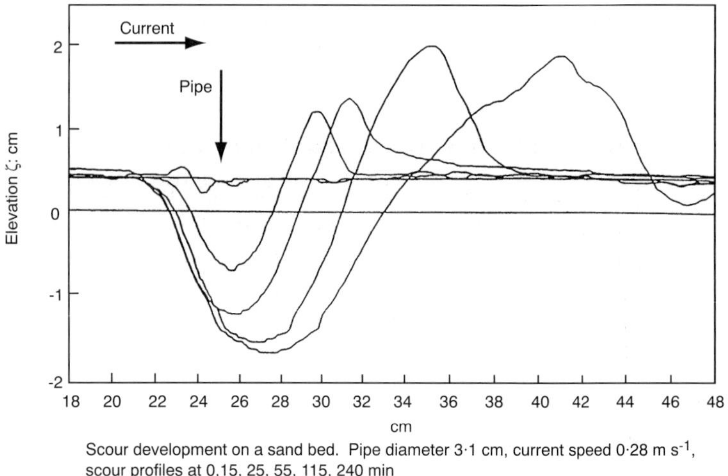

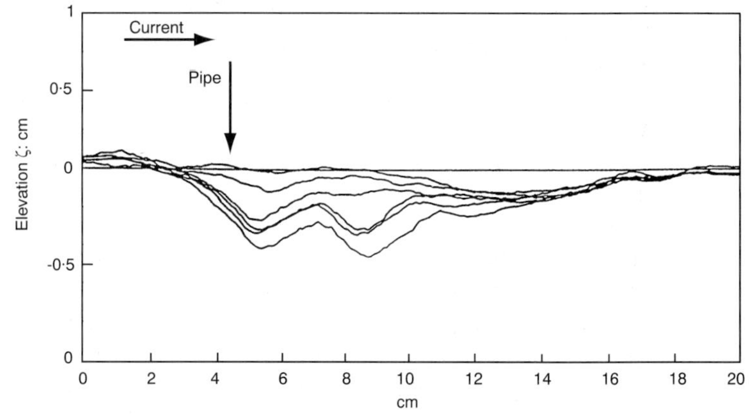

Figure 31. Influence of cohesive sediment on pipe scour (reproduced from Pluim-van der Velden and Bijker, 1992, by permission of Bentham Press)

(sand mixed with kaolinite) eroded only very slowly and the scour hole was smaller than for the case of pure sand. With the high cohesion bed (natural silt–sand mixture) the velocity required to induce the erosion was much higher and the sediment eroded in lumps leading to an irregular scour profile and no deposition downstream (Figure 31). At very high velocities the equilbrium scour was still able to develop fully to a depth of $0{\cdot}6D$.

7.4.12. *Orientation to wave–current flow*

Unpublished wave flume studies at HR Wallingford have indicated a significant reduction in the scour depth beneath a pipeline sitting on the sea bed in angled wave attack. The scour depth beneath the pipeline oriented at 45° to the wave crests was only 75% of the depth when the wave crests were parallel to the pipeline.

7.4.13. *Trenching and trench infill*

Laboratory experiments on the interaction of the flow field with a pipeline sitting in the bottom of a v-shaped trench have been reported by Fredsøe (1978). He placed a 74 mm diameter D pipeline in a trench cut into a sand bed at 45° to the oncoming wave crests. Two trench side slopes were tested with slopes $1V{:}3H$ and $1V{:}6H$ and the trench depth was equal to $1{\cdot}9D$. The experiments showed that no scour was observed when the pipeline was placed in the trench rather than on the sea bed and the wave induced bedload transport was able to infill the trench providing additional protection for the pipeline.

Wider trenches with side slopes of greater than $1V{:}8H$ may have less sediment trapping capacity and are possibly less effective in preventing scouring under the pipeline.

7.4.14. *Effect of storms*

A combined experimental and numerical study of the scouring and backfilling of pipelines by waves during a changing wave climate has been reported by Fredsøe *et al.* (1992). The predicted changes in scour depth after the transition from one wave condition to another compared satisfactorily with the measurements. The equilibrium scour depth was predicted using Equation (44) with the value of KC after the change in wave conditions and the rate of change (i.e. time-scale T^*, Equation (45)) by the bed shear stress after transition θ_{final}. The time-scale was also dependent upon KC both prior to and after the transition: the smaller the change in KC the smaller is the value of T^* for a given θ_{final}.

7.4.15. *Intertidal pipelines*

Observations of the scour depth under fixed outfall pipes running across intertidal areas show the maximum scour depth to be commonly of order $0{\cdot}6D$. Larger total scour depths (or burial) are caused by general changes in the level of the bed or beach.

The results of pipeline tests at intertidal locations in the Taw-Torridge estuary are reported by HR Wallingford (1972). Two sites were chosen, a flat bed site ($d_{50} = 0{\cdot}17$ mm) and a site with mobile sandwaves ($d_{50} = 0{\cdot}2$ mm, sandwave height 1 to 2 m and wavelength 20 m). During spring tides the water depth was around 3 m and the current speed perpendicular to the pipelines exceeded $1\,\mathrm{m\,s^{-1}}$. A selection of pipes, concrete coated steel (OD = 0·61 m, 0·406 m and 0·219 m) and flexible armoured (OD = 0·157 m), were tested with different submerged weights. Water depth effects on the pipeline–bed interaction were generally considered to be insignificant. The results from the tests with the 0·61 m and 0·157 m pipes are summarised below.

The 92 m length of the 0·61 m diameter pipe was installed (with end anchors to represent the rigidity of a continuous pipeline) at high water and perpendicular to the tidal flow. On the following low tide the pipe was observed to be lying in a shallow scour trench with undercutting over a greater part of its length. The scour profile upstream was almost at the angle of repose of the sand and the downstream slope was more gentle. The pipe was flooded to a submerged weight of $352\,\mathrm{kg\,m^{-1}}$ and was observed to settle into the bed over the period of a few days resting in a wide gently sloping scour pit (sideslopes $1V{:}7H$). The width of the scour pit expanded at the spring tide stage (Figure 32*a*, maximum extent and depth of scour) and contracted at neap tidal stage (Figure 32*b*). The maximum scour occurred after about 4 days with an associated tidal current speed adjacent to the bed of $1\,\mathrm{m\,s^{-1}}$.

The vertical settling of the pipe with respect to a fixed datum was observed to be quite uniform along the length; the average

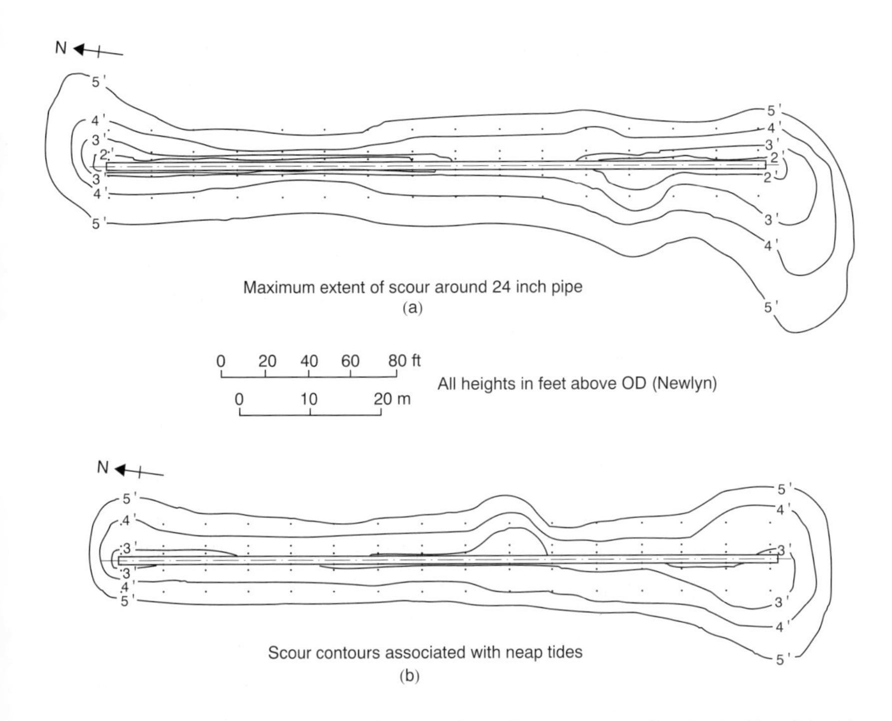

Figure 32. Scour development around a 24 inch steel–concrete pipeline in the Taw-Torridge estuary (from HR Wallingford, 1972)

settlement for the central 30 m length was 0·36 m or 0·6D in 5 days after which time the pipe stabilised. After 6 days the extent of the scour area decreased as the tidal range (current velocity) decreased in magnitude towards the neap range presumably due to an equilibrium between scouring and the infill of material transported by the current. The end effects depicted in Figure 32 were not found to influence the overall pipe behaviour. Over the next three weeks the scour pit continued to encroach on the pipe and eventually completely covered it with a depth of about 0·3 to 0·4 m of sand over the top of the pipe. For the remainder of the survey period the pipe remained buried under a rippled sand bed.

The 114 m length of 157 mm OD flexible polyethylene pipe was laid with some plan curvature approximately perpendicular to the crestline of the sandwave field. The pipe followed the bed topography quite well except for the occurrence of short free-spans at sharp changes of slope. The scour/burial and vertical displacement of the pipeline was monitored as well as the bed topography. The surveys indicated that the pipeline initially lay proud of the bed surface with scour burial taking place at the sandwave crests and deeper burial being caused by sandwave migration. At one position the vertical displacement of the pipe was measured as $2{\cdot}9D$ or 0·45 m over a period of 15 days after the initial lay and the pipe was covered by up to 0·3 m of sand in places. The burial depth and percentage of pipeline buried varied with changes in the position of individual sandwaves. The combination of the severe bed conditions and maximum current speeds exceeding $1\,\mathrm{m\,s^{-1}}$ indicated the usefulness of flexible pipelines for unprepared sites with significant surface topography.

7.4.16. *Scour prediction methods for pipelines—3-dimensional case: span length*

Normally the scour pattern is *not* uniform along the pipeline. In a situation where the pipeline is suspended over a scour hole the final scour pattern depends on the speed at which the scour hole spreads along the pipeline: if the hole is spreading slowly the sagging of the pipe will occur slowly and the situation is almost static whereas if the development of the scour hole along the pipeline is rapid the pipe will move relatively fast towards the bed and this sagging velocity has been shown by Fredsøe *et al.* (1988) to have an effect on the equilibrium scour depth. The exact influence on the scour depth depends upon the sagging velocity. Hansen *et al.* (1986) have investigated numerically the scouring at pipeline free-spans and in the shoulder area where the pipeline is supported in the sand bed. Leeuwenstein *et al.* (1985) deduced from laboratory tests of 3-dimensional scour that the propagation velocity of the scouring of the supporting ridges of sand takes place at between 10^{-2} and 10^{-1} metres per hour. Sumer and Fredsøe (1994) presented a simple formula based on soil mechanics for predicting the critical shoulder length required to support a pipeline of a given density. For shoulder lengths shorter than the critical value the pipeline begins to sink owing to soil failure.

Lucassen (1984) postulated that scour of one to two pipe diameters was possible given the combined effects of pipeline sagging and two-way tidal flow with wide scour-trenches.

In the situation with no or slow sagging it is useful to determine the maximum free span of the pipeline which can develop: this is of course dependent on material properties of the pipeline. An important parameter is the stiffness length (Fredsøe *et al.*, 1988)

$$L_s = \sqrt[3]{\frac{EI}{m}} \tag{46}$$

where E = modulus of elasticity for the pipe ($\mathrm{N\,m^{-2}}$), I = moment of inertia for the cross-section of the pipe ($\mathrm{m^4}$) and m = the weight of the pipe in the fluid per unit length ($\mathrm{N\,m^{-1}}$). Fredsøe *et al.* (1988) found that the maximum span length could be determined as

$$L_{max} = 3{\cdot}35 D^{1/4} L_s^{3/4} \tag{47}$$

where L_s is given by Equation (46). Sagging associated with free-span generation will cause additional erosion. Fredsøe *et al.* (1988) found that a rough estimate of the equilibrium scour depth in the case of a sagging pipe was one pipe diameter.

7.4.17. Effect of pipe vibration

Flow past a pipeline can under certain conditions (mainly depending on the Reynolds number) cause shedding of vortices from the trailing side of the pipe (lee side scour). The vortex shedding can, in turn, result in vibrations of the pipe itself. Mao (1986) investigated the effects of the flow induced vibration of a free-spanning pipeline on the equilibrium scour depth under the pipe. He found that for a range of initial gaps beneath the pipe the equilibrium scour depth was increased by vibration especially when the free moving pipe was able to impact the sea bed. Cross-flow vibrations of pipelines has been shown to increase the equilibrium scour depth by 20%–50% (Sumer *et al.*, 1988b). It is therefore important to make a rough assessment of the likelihood of cross-flow induced pipeline vibrations.

7.5. LARGE VOLUME STRUCTURES

The potential for scour around large volume structures needs to be considered during both the installation and operational phases of the lifetime of a structure. For example, the process of installing a gravity base structure offshore (O'Riordan and Clare, 1990) results in an ever decreasing gap between the lowermost edge of the structure and the sea bed for the tidal flow to pass through. This reduction in gap increases the potential for scour to occur, even if it is only for a short period of time. Post-installation scour is important because of the eccentric foundation loading that can result.

There is only a limited amount of experimental data and numerical studies of the flow field and the scouring around gravity installations. Most studies have been structure and site specific for a client and therefore the information obtained in such studies often remains unreported in the open literature. Dahlberg (1981) presents a large amount of useful information based on field experience at a number of installations, primarily in the North Sea. Also, a relatively large amount of work has been reported on scour at large cylinders: this is relevant for gravity installations such as the Ekofisk oil storage tank (dimensions 93 m diameter and 90 m tall in 70 m water; Lee and Focht, 1975).

In coastal and estuarine areas the scour around coffer dams and artificial islands can be considered in the same class of large volume structures.

7.5.1. *Water depth*

When the relative cylinder diameter is large ($D/h > 0{\cdot}5$) the scour pattern is different from the slender pile case and in wave motion the additional effect of wave diffraction begins to become important when the cylinder diameter to wavelength ratio D/L exceeds a value of 0·2 (Rance, 1980). For typical wave periods in the North Sea of 4 to 8 seconds (Draper, 1991) the corresponding deep water wavelengths L_0 are 25 m and 100 m and hence diffraction effects will become important for structures with diameters larger than 5 m and 20 m respectively. For $D/L \geq 1$ the situation becomes dominated by wave reflection. As the caissons of gravity base structures are relatively large, the effects of wave diffraction and reflection need to be considered. The processes in deep water ($h/L > 1/4$) will be complicated because of wave pressure attenuation effects and the influence of the slab-sides on the wave-induced flow field (Stubbs, 1975).

7.5.2. *Scaling of scour depth*

The scour depth for large cylinders does not scale with the diameter in the same fashion as the scour that develops around a slender pile, and scour and siltation can occur simultaneously around the periphery of the cylinder (Rance, 1980; Katsui and Toue, 1992). Physical model results at HR Wallingford (Rance, 1980) for the scour around a large circular cylinder indicated maximum scour depths of $0{\cdot}032D$ for waves and $0{\cdot}064D$ for collinear waves and currents, plus accretions of $0{\cdot}028D$ in some areas adjacent to the installation (Figures 33 and 34).

7.5.3. *Scour depth and position*

Physical model tests to determine the scour around large cylinders (Torsethaugen, 1975) indicate that the maximum scour depth in a steady current does not occur at the upstream face of the cylinder as in the slender pile case but at about 45° from the axis of the oncoming flow (Figure 10). The maximum scour depths are dependent upon both the ratio of water depth to pile diameter and the actual pile diameter. For live-bed scour with $h/D = 0{\cdot}66$ the scour depths for a 0·75 m diameter pile and 0·1–0·15 m diameter piles were about $0{\cdot}2D$ and $1D$ respectively.

May and Willoughby (1990) measured the equilibrium scour development around large cylinders and square piles facing into the flow (e.g. Figure 35 for clear-water scour). The results of their tests indicated that the scour depth for the square piles was generally a maximum at the corners, but the data indicated that the scour depth was also a function of the pile size and relative water depth, in agreement with the data of Torsethaugen (1975). Both these sets of experiments determined the maximum equilibrium scour depths that would occur after long exposure to a steady unidirectional flow. May and Willoughby measured scour depths significantly smaller than predicted by the Breusers *et al.* formula (Equation (35)) in the range $0{\cdot}1 < h/D < 3$, and

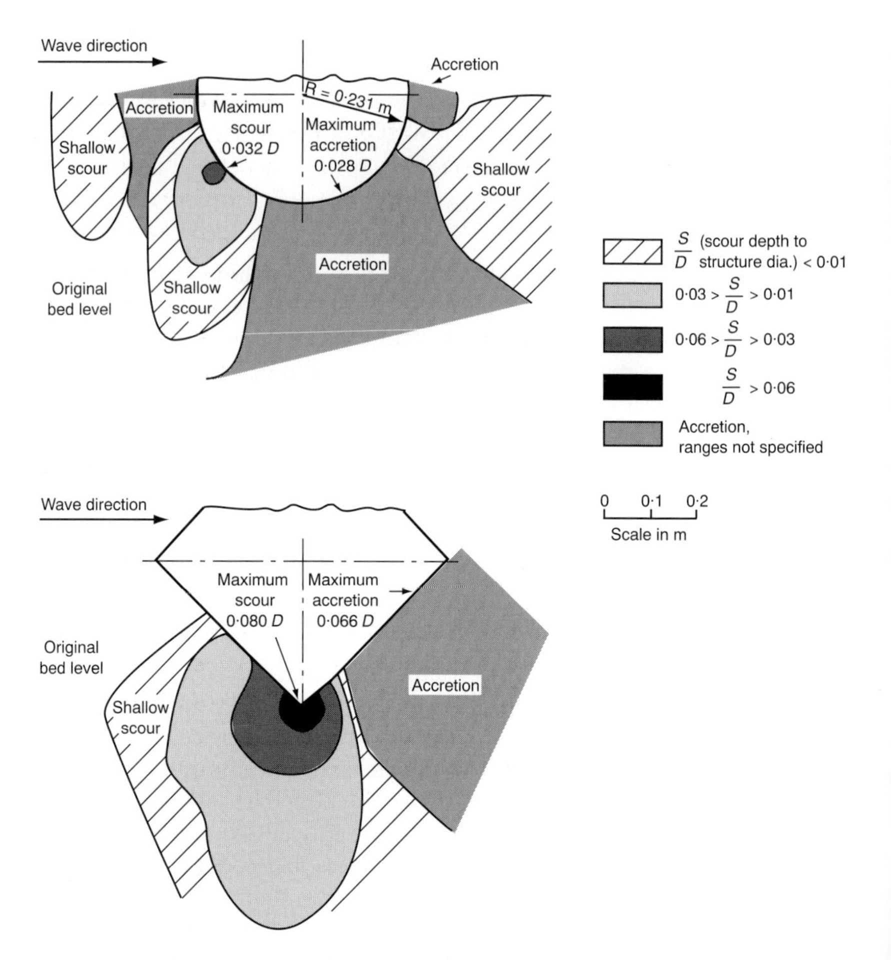

Figure 33. Laboratory data for scour development around large structures in waves (reproduced from Rance, 1980, by permission of the Society for Underwater Technology, London)

based on these results presented a revised equation for shallow water effects. Following the study of May and Willoughby further work on scour in tidal flow is being performed by HR Wallingford at the present time.

Breusers (1972) has undertaken tests of a gravity structure with rectangular rafts at a geometric scale of 1:50 and reported prototype equivalent scour depths of 6 m to 9 m. Indicative scour

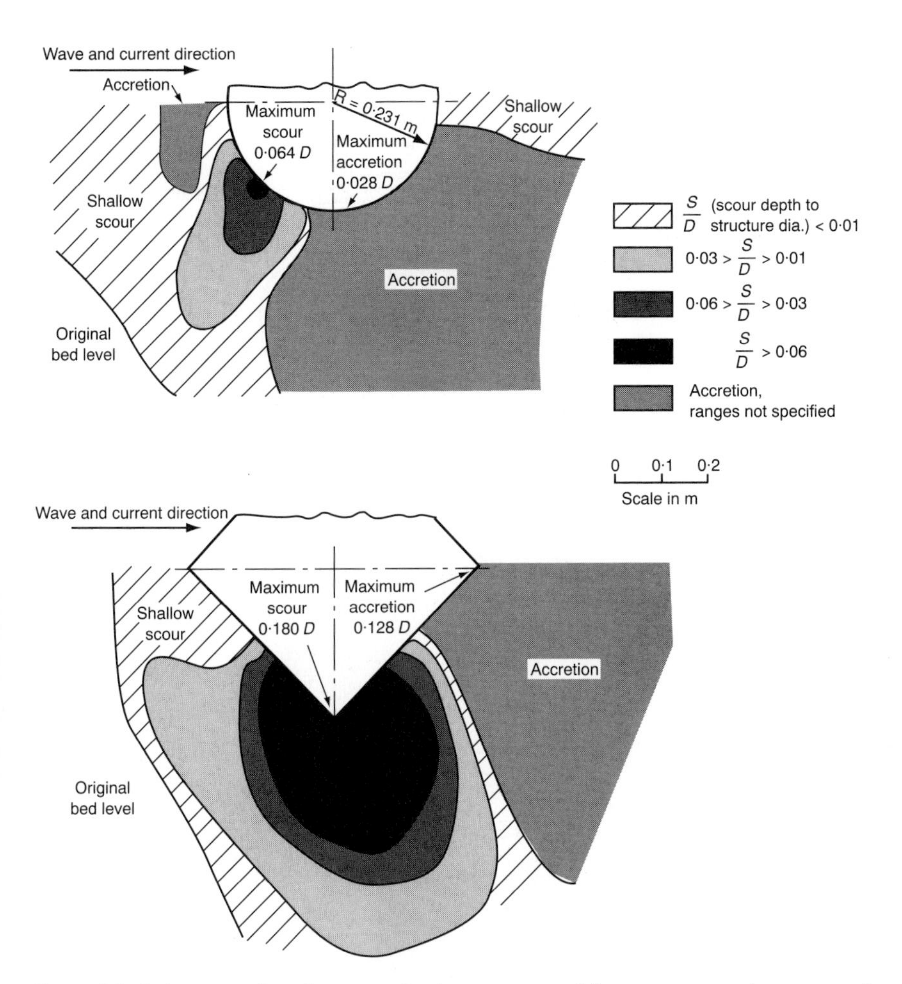

Figure 34. Laboratory date for scour development around large structures in waves and current (reproduced from Rance, 1980, by permission of the Society for Underwater Technology, London)

tests around similar structures in a steady current have been reported (Ninomiya *et al.*, 1972). Observations of the post-installation scour depths at gravity structures in the North Sea have been collated by Dahlberg (1981) and a maximum scour depth of around 2 m occurred at the corners of some of the structures. These findings compare well with the numerical simulation by O'Riordan and Clare (1990). Other potentially

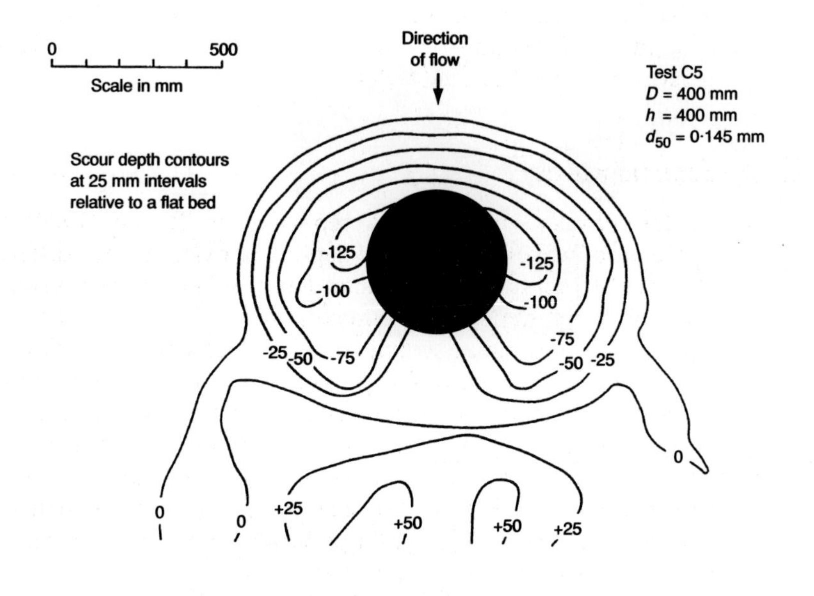

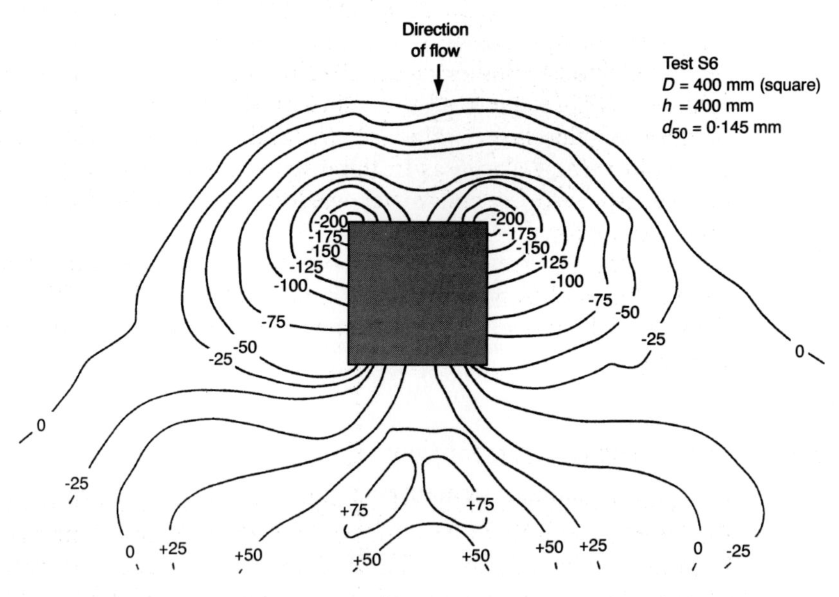

Figure 35. Typical scour development around large structures in a steady flow, clear-water scour $U/U_{cr} = 0{\cdot}61$ (from May and Willoughby, 1990)

vulnerable areas are at the leading face of the installation and around templates and pipelines associated with the installation (O'Riordan and Clare, 1990).

7.5.4. *Scour in waves*

In the physical model tests reported by Rance (1980) the influence of waves on the scour depth was examined for different shapes of structure, using a water depth of 0·5 m and Bakelite as the mobile bed material. The maximum live bed scour depths at square and hexagonal structures with different orientations to the wave direction were a few per cent larger than for the circular cylinder, for the one condition tested. The scour pits extended to about half a diameter from the edge of the structure for the circular cylinder and to a distance equal to one 'diameter' for the square structure (Figure 33). The interaction of the structure with the wave field produced a pattern of erosion and accretion upstream of the cylinder.

It should be pointed out, similar to the steady current case, that it is not appropriate to extend the predictive formula for slender cylinders (Sumer *et al.*, 1992b) to the case of very large cylinders. This is because the value of KC will be very small, because of the large value of D, and probably below the threshold value KC = 6 as in the tests by Rance. Consequently the horseshoe vortex and vortex shedding are not the dominant processes as for the slender pile.

The effect of breaking waves on the scour development at large cylinders has not been determined but in the limit $D/L \gg 1$ the wave field and scour development could approximate the sea wall case.

7.5.5. *Scour in wave–current flow*

In the tests of Rance (1980) the equilibrium live-bed scour depth for wave–current flows was deeper than for the wave-alone case, but only one wave and current flow condition was tested (ratio of peak oscillatory flow velocity to unidirectional flow = 2·5) (Figure 34). The extent of scour was about $1D$ from the edge of the structure. Experimental data and numerical modelling of the scour evolution for a similar case have been published by Katsui and Toue (1992) and although the general features of the scour topography related to the wave diffraction and reflection were predicted the detailed patterns of erosion and accretion around the structure were not.

Bishop (1980) reported the results from scour measurements made around the base of the first Christchurch Bay tower (south coast of UK), a gravity structure without a bed penetrating skirt. Scour depths in the sandy sediment were typically 0·5 m to 1 m at the periphery of the base with a saucer-like depression extending a distance of between 12 m and 20 m from the base. The water depth was 9 m and the base diameter 10·5 m. Scour was also apparent with the second tower even though a skirt had been introduced. The area of sea bed that required repair lay in the sector E to SE in the region between the axis of the tidal current and wave direction. Observations by Maidl and Stein (1981) at the Nordsee research platform (southern North Sea) showed that local scouring up to 0·8 m deep could form at the unprotected cutting edges of the platform base and largely resulted from exposure to the tidal current. This platform has an octagonal base 75 m in diameter and 4·5 m in height. Dahlberg (1981) observed scour depths of 2 m in dense fine sand at two corners of the square-based Frigg TP1 platform (square sides of length 72 m in 104 m water depth, northern North Sea).

Also of interest are the results of flume tests performed to assess the scour around the piers for the western bridge of the Great Belt link in Denmark (Hebsgaard *et al.*, 1994). The prototype bridge structure comprises a number of rectangular bottom caissons of height 7·5 m to 9 m with dimensions of 17 m by 30 m for the bottom part and 6·4 m by 30 m for the upper part. The occurrence of scour in the physical model tests was attributed to the shear stress amplification that occurred adjacent to and downstream from the corners of the structure in the direction of the wave and current flow. The maximum scour depth beyond the protection layer was 3 m at a distance of about 20 m to 25 m from the caisson. The depth and position of scouring varied with the changes in wave–current direction. Scouring of the protective gravel blanket surrounding the caisson was also observed.

Although large coffer dams are frequently used in coastal construction projects it appears that no information exists on the scouring around these installations by near-normal incidence waves and cross-currents.

7.5.6. *Time development of scour*

The observations from Christchurch Bay and the North Sea referred to in this Section suggest that scour effects can manifest themselves rapidly if the installation has been subjected to sufficiently strong wave–current conditions.

O'Riordan and Clare (1990) computed that 2 m of scour could occur at the corners of a rectangular gravity structure within 16 hours of touchdown on spring tide conditions (maximum ambient velocity $0{\cdot}5\,\mathrm{m\,s^{-1}}$) but only to about 0·2 m on neap tides (velocity $0{\cdot}3\,\mathrm{m\,s^{-1}}$).

May and Willoughby (1990) have reported comprehensive laboratory measurements for the time development of scour around large volume structures. These results are currently being used in research at HR Wallingford to develop a scour predictor for tidal flow.

7.5.7. *Shape*

Dahlberg (1981) concluded from field observations that the gravity structure Frigg TP1—72 m square based installation—was more sensitive to scour than the more or less circular base of the nearby Frigg TCP 2. A square foundation is much more sensitive to scour than a circular one. The effect of cross-sectional shape has been examined by May and Willoughby (1990) who found that the scour depth for large square shaped structures was on average 1·3 times larger than for an equivalent cylindrical structure.

The scour potential can be evaluated based upon a consideration of the available data. Alternatively the potential for scour can be determined by flow modelling (e.g. O'Riordan and Clare, 1980) or measurements of the shear stress amplification around a structure (Hebsgaard *et al.*, 1994). Model tests or hybrid physical–numerical modelling are recommended to assess complex geometries.

7.5.8. *Angle of attack*

The velocity amplification around a square caisson varies with the angle of attack (O'Riordan and Clare, 1990; Hebsgaard *et al.*, 1994) and therefore the scour and backfill processes will vary their position with time as the direction of the tidal current and waves vary. Rance (1980) has presented scour experiments which illustrate the influence of angle of attack (e.g. Figures 33 and 34).

Some of the perceived advantages to be gained by streamlining an installation will be lost in the open sea because of the continually varying wave–current climate (magnitude and direction). However, it should be possible to orientate the structure and reduce the angularity of its design to mitigate the scouring effect due to tidal and storm surge events, the predominant direction of which will usually be known.

7.5.9. *Scour due to motion of structure*

In storms the rocking motion of gravity structures has been observed at the North Sea installation Frigg CDP1 (Dahlberg, 1981) and in an extreme case at the Christchurch Bay research tower as a result of the undermining of the unskirted foundations of the first tower during a series of severe storms (Bishop, 1980). The soil resistance to overturning moments gives rise to a rocking motion which can lead to erosion arising from the flows under the installation and the pressure gradients set up in the bed. Water will flow preferentially along lines of weakness in the fabric of the soil and when this takes place a localised form of scour called *piping* can occur.

7.5.10. *Effect of resistant layer*

The presence of a resistant clay layer may inhibit the scour depth if its thickness is less than the potential scour depth for the installation. Gravity structures are usually installed with skirts that penetrate into the bed to provide protection against underscour. However, at the Christchurch Bay tower the clay substrate was also eroded beneath the skirt probably due to the mechanical disturbance caused to the soil by the presence of the skirt moving under the rocking motion (Bishop, 1980).

7.5.11. *Seasonality*

Although the scour depth is expected to be greatest during the winter months Dahlberg (1981) reported on the occurence of 2 m deep scour pits at two corners of the Frigg TP1 structure even during the summer months.

7.6. SEA WALLS

It is important to be able to predict the interaction between the wave field incident on a sea wall and the adjacent sea bed when designing a structure to withstand storm waves.

The scour development at the toe of the sea wall is controlled by the near-bed orbital velocities generated under the combined influence of incident and reflected waves. In addition significant pressure gradients can be generated in the bed immediately seawards of the wall and these can be effective in disturbing the bed sediments. Müller (1995) reports experimental results which demonstrate large sea bed pressures and high pressure gradients at the bed, arising from wave impact pressures being transmitted down through the water column to the bed and away from the wall. His results indicate possibly lower absolute pressures but considerably steeper gradients acting away from the sea wall than some previous work. As beach profiles vary temporally and especially seasonally, those sea walls fronted by intertidal beaches will also be subject to changes of the general bed level. In practice the level of sediment at the toe of the sea wall will be controlled by natural variability of the beach profile due to subtidal and intertidal processes, as well as the local scour mechanisms produced through the presence of the sea wall itself. It may often be difficult to identify and decouple the two.

The effect of local scour which occurs near the foot of the sea wall (defined as *toe scour*) is threefold (Oumeraci, 1994b) as it may lead to:

(*a*) a gradual dislocation of the rubble mound or blockwork foundation,
(*b*) a decrease in the geotechnical stability of the structure,
(*c*) a modification of the wave and flow conditions in front of the structure.

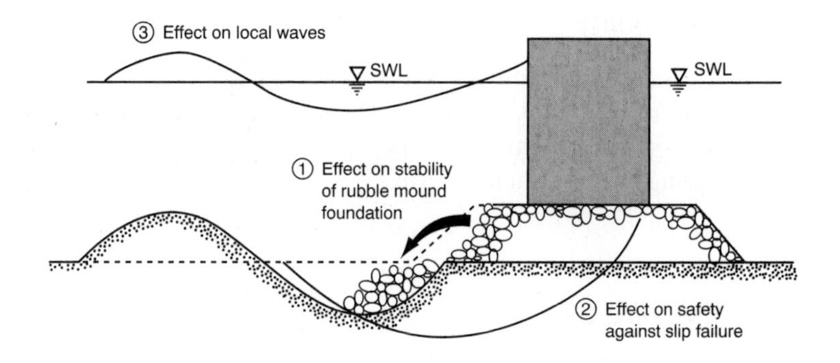

Figure 36. The effects of toe scour at sea walls (reproduced from Oumeraci, 1994b, by permission of the Port and Harbour Research Institute, Yokohama)

These effects are indicated in Figure 36 taken from Oumeraci (1994b). As a result of (*c*) the area of sea bed or beach in front of the structure may respond in a very different fashion from what would be expected under the incident conditions alone.

There are a number of approaches to considering toe scour at sea walls, the most commonly used being rules-of-thumb or other empirical correlations based upon laboratory and field observations, or scaled-down physical models. The scour depth at sea walls is usually non-dimensionalised with an appropriate measure of the wave height, e.g. significant wave height offshore.

7.6.1. Scour development and associated hydrodynamics

For vertical sea walls the maximum toe scour depth is approximately equal to the deepwater wave height (Fowler, 1993, wave steepnesses 0·003 to 0·036, fine sand) or the incident unbroken wave height H (Powell, 1987, wave steepnesses 0·02 to 0·04, shingle). The maximum scour depth occurs when the sea wall is located in an initial water depth of $1{\cdot}5H$ to $2{\cdot}0H$ (Powell, 1987; CIRIA, 1996) because there is a dependency of the scour depth both on the initial water depth at the wall and on how close to equilibrium the beach profile is prior to the onset of scouring. Alternatively (Fowler, 1993), the maximum scour depth occurs at a distance from the sea wall of between $0{\cdot}5 \leq x_{sw} \leq 0{\cdot}67$, where $x_{sw} = x/x_b$ is the ratio of the distance of the sea wall from the point of wave breaking x divided by x_b the distance of the point of wave breaking from the intersection of mean sea level with the pre-scour beach profile.

Recent experimental work (CIRIA, 1996, after Powell and Lowe, 1994) supports the rule of thumb that in coarse sediment (shingle, $5\,\mathrm{mm} < d_{50} < 30\,\mathrm{mm}$) the maximum scour is approximately equal to the incident unbroken significant wave height H_s, for sea steepnesses in the range $0{\cdot}02 < H_s/L_m < 0{\cdot}04$, where L_m is the mean wavelength. Figure 37*a* is a contour plot of the ratio of the dimensionless toe scour S_{3000} and accretion at vertical walls after 3000 waves divided by the incident wave height H_s. This diagram was prepared from laboratory flume data with irregular waves and a single sediment grading. The magnitude and pattern of S_{3000}/H_s will vary with changes in the initial water depth at the toe of the sea wall, sediment size, wave height and period, number of waves, and sea wall profile and surface roughness. The results in Figure 37*a* can be adjusted for a different number of waves with results obtained for the time development of scour (see Figure 39).

New research by Carpenter and Powell (1998) and Powell and Whitehouse (1988) has resulted in the production of a toe scour contour plot for sand sized material (Figure 37*b*, sediment $d_{50} = 0{\cdot}200\,\mathrm{mm}$). The figure depicts the same axes as Figure 37*a*, although the ranges covered are somewhat different. The data used to construct this figure were derived from a large number of simulations with the HR Wallingford cross-shore process based numerical model COSMOS-2D. The model had been validated against the large-scale SUPERTANK measurements of scour in front of a vertical sea wall (McDougal *et al.*, 1996). Irregular waves were used in the simulations and an initial beach slope of $1V{:}75H$ was specified which corresponds to fine sand (CIRIA, 1996). Owing to the limited capability of cross-shore models to predict the onshore movement of sediment on beaches the accretion phase of Figure 37*b* has not been plotted.

The patterns of the scour contours on Figures 37*a* and 37*b* are broadly similar. In both cases the maximum scour depth at a wave steepness (H_s/L_m) of 0·01 is $1{\cdot}5H_s$ for an initial toe water depth of $2{\cdot}0H_s$. However, the rate of change in the scour depth contours with water depth for low wave steepness is more rapid in shingle than in sand, whereas for higher wave steepness the converse is true. A significant difference for the results with sand

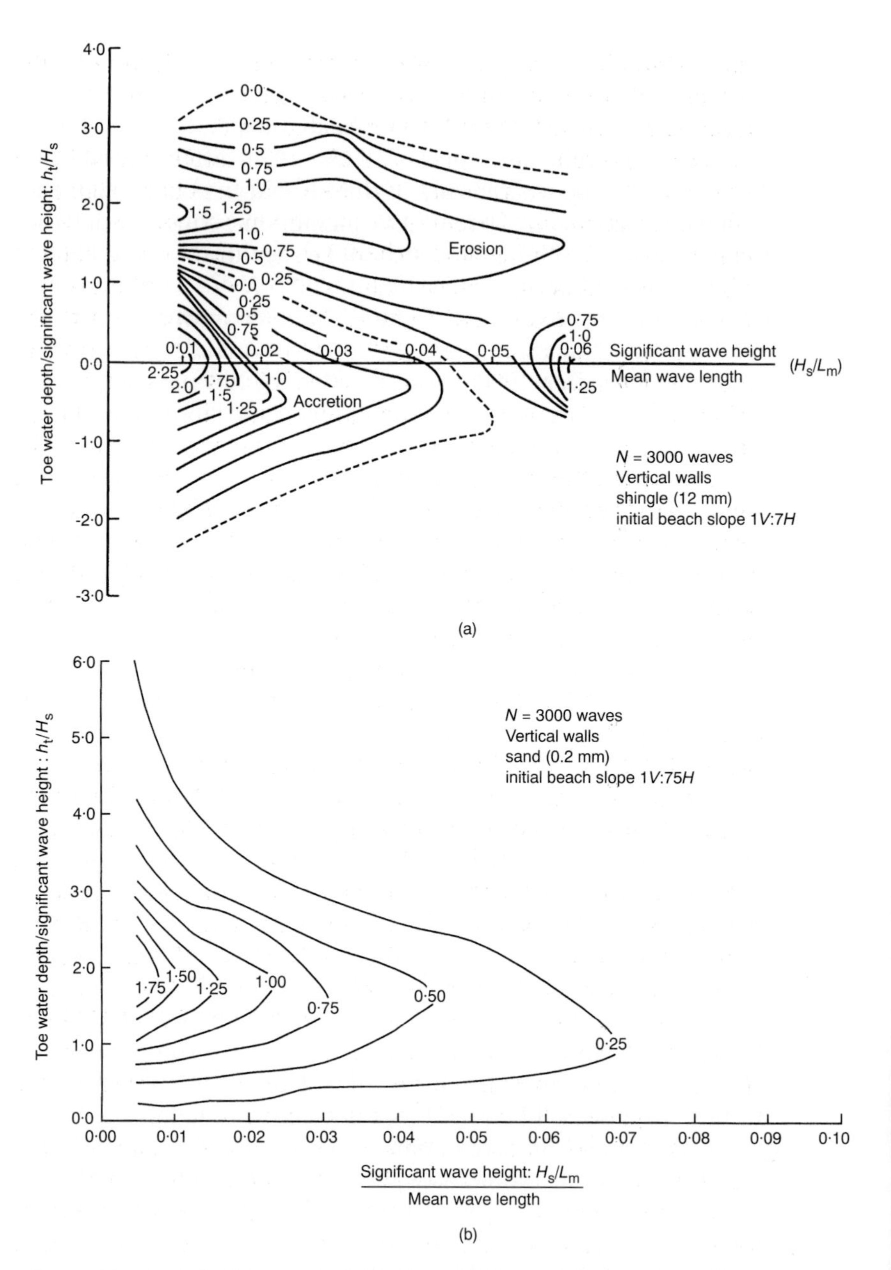

Figure 37. Scour development in front of a vertical sea wall: contours of dimensionless toe scour S_{3000}/H_s in relation to dimensionless water depth and wave steepness: (a) shingle-sized material reproduced with permission from Powell and Lowe, 1994, (b) sand-sized material (from Carpenter and Powell, 1998)

is the much smaller range of wave steepness over which the scour depth is equivalent to the incident wave height.

Fowler (1993) has reported field results from sandy sites which support the premiss that $S_{max}/H_s \leq 1$ even in hurricane storm conditions. In additon, Kraus and McDougal (1996) refer to observations made on the US east coast where surveys taken over a period of seven years including before and after major storms recorded that a scour trough was never observed in front of any of the sea walls studied. However, the generality of this finding is difficult to assess as it is clear the scour development is dependent upon initial conditions of beach slope and water level at the wall (Fowler, 1993; CIRIA, 1996) as well as sediment size.

The amount of wave reflection has been shown to exert an influence on the bed scour profile of a horizontal sea bed in front of a vertical wall (Oumeraci, 1994b), and the pattern of scour is different for relatively fine sand ($U_{wmax}/w_s \geq 10$; Irie and Nadaoka, 1984) and for relatively coarse sand ($U_{wmax}/w_s < 10$); here U_{wmax} is the maximum amplitude of the wave bottom orbital velocity and w_s is the settling velocity of the bed sediment. For fine sand in suspension the orbital velocity and mass transport velocity in the bottom boundary layer produced by the standing wave field causes scour to occur at the nodes (a distance of $L/4$ and $3L/4$ from the wall) of the standing wave field (for regular period waves) and deposition at the antinodes (i.e. at the wall, $L = 0$, and $L/2$—Figure 38). For coarse sand the sediment is eroded at distances of $L/8$ and $3L/8$ from the wall and deposited at the nodes and antinodes (Figure 38). The scour-deposition pattern for both regular and irregular incident waves is similar within a distance $L/2$ from the wall although beyond this the scour pattern becomes less pronounced under the action of irregular waves.

For typical prototype conditions Oumeraci (1994b) concluded that transport by suspension of relatively fine sand was likely to be the commonly occurring mode of transport, leading to one deep scour hole per half wavelength, as opposed to two less deep ones per half wavelength for coarse sand. Typical scour profiles for shingle are shown in Figure 39.

It is clear that the waves reflecting from a sea wall exert a strong controlling influence on the wave kinematics and the magnitude and pattern (location) of bottom orbital velocities in

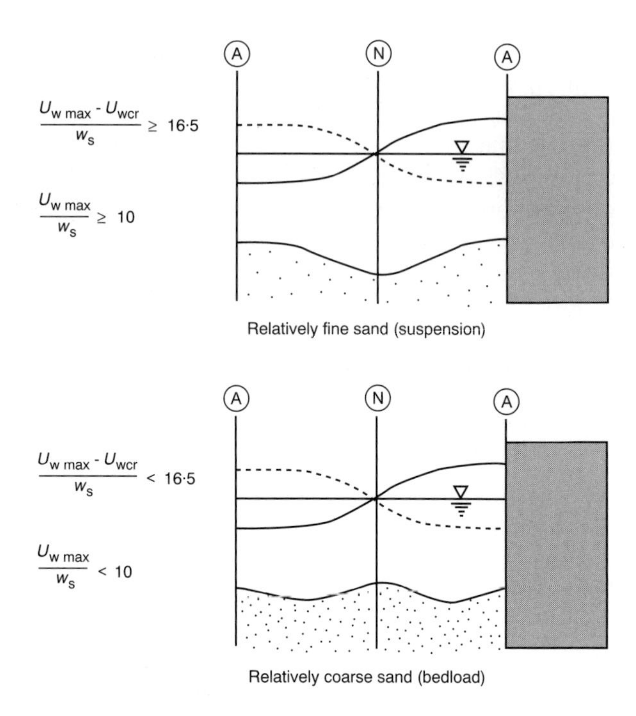

Figure 38. Scour profile in front of a vertical wall (reproduced from Oumeraci, 1994, by permission of the Port and Harbour Research Institute, Yokosuka)

front of the sea wall (O'Donoghue and Goldsworthy, 1995). O'Donoghue and Goldsworthy (1995) report data for the bottom orbital velocity from 2-dimensional flume experiments under irregular waves on beach slopes of $1V{:}7H$ and $1V{:}40H$ and closely predict these values using linear wave theory (after Soulsby, 1987—also see Soulsby, 1997). This approach yielded encouraging results when predicting the bottom velocity for the same beaches fronted by a smooth vertical, 45° or 30° (to the horizontal) sea wall, providing the wave reflection characteristics were adequately specified. However, the significance of wave reflection on toe scour at sea walls fronted by beaches has been questioned recently by McDougal *et al.* (1996).

The influence on scour of waves breaking in front of the sea wall is the subject of ongoing research as the additional

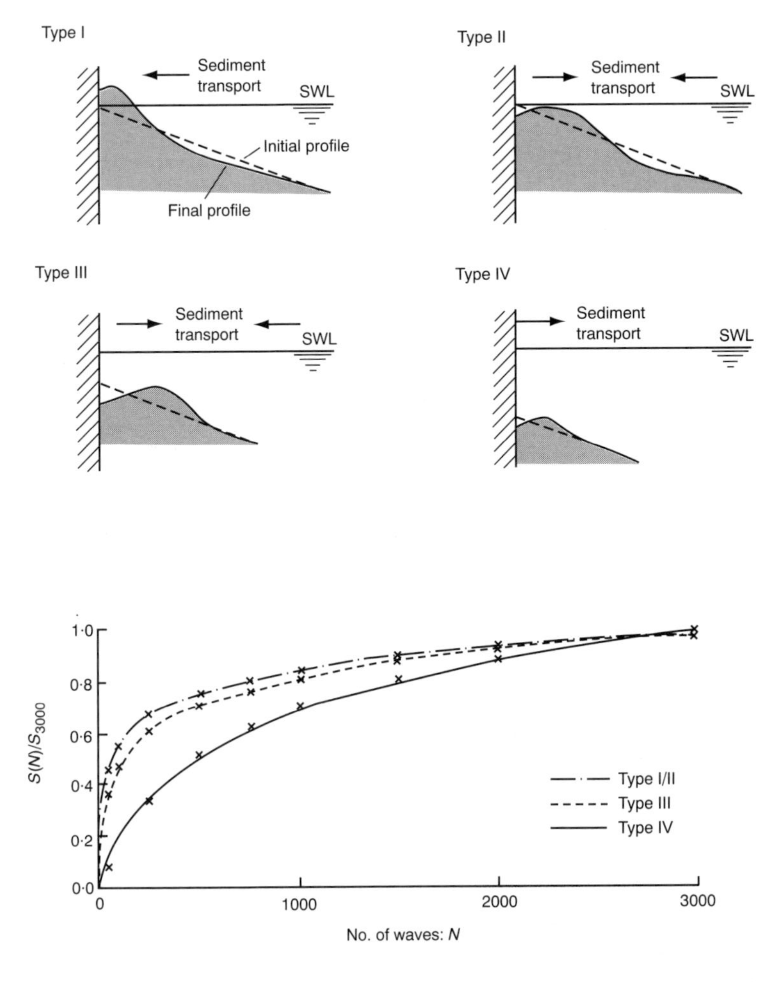

Figure 39. The influence of initial beach level on scour response in shingle (reproduced with permission from Powell and Lowe, 1994)

turbulence generation may be significant in controlling the scour process and ultimate scour depth.

7.6.2. Time development of scour

A number of experiments have been conducted to investigate how the toe scour develops through time (Herbich *et al.*, 1984;

CIRIA, 1996; McDougal *et al.*, 1996). McDougal *et al.* (1996) state that scour depth relationships are only applicable to specific initial conditions. The data of Powell and Lowe (1994) illustrate well the typical time development (Figure 39) and shows that the initial scour response at the wall can be very different for different initial conditions. These data are specifically for shingle but are reasonably typical for scour at sea walls in general, initially rapid scouring followed by a period of slower development towards an equilibrium value. The time development plots in Figure 39*a* can be used to correct the contoured scour depths in Figure 37 for events of shorter duration than 3000 waves. The curves in Figure 39 indicate that S_{3000} is more or less the equilibrium value. A similar type of approach for sand sized material is described in Carpenter and Powell (1998), also Powell and Whitehouse (1988). The time development was slower (10 000 waves) in the case of sand than was found for the shingle sized materials.

McDougal *et al.* (1996) tested a numerical cross-shore sand transport model against data from the large scale SUPERTANK experiments with a vertical sea wall. Although the agreement between model and test data was encouraging the response adjacent to the wall was not always closely predicted. Despite this they were able to use the numerical results to investigate the time evolution of the scour and derived an approximate negative exponential dependency, essentially similar to the approach used above for vertical piles and for pipelines (Equation (3)):

$$S(t) = S_e\,(1 - \exp^{-\mu \frac{t}{T}}) \tag{48}$$

here $S(t)$ is the toe scour depth at time t, S_e is the equilibrium toe scour depth, and T is the wave period. A best fit line to their (numerical) data gave $\mu = 0{\cdot}000321$. From a comparison of this time development equation with the data in Figure 39 it appears that there is probably a dependency of μ on the initial bed condition/water depth at the wall and the sediment transport rate. The shape of the scour development curve also exhibits some as yet undefined dependency on the physical conditions. To account for such effects would effectively involve replacing

$$\left(-\mu \frac{t}{T}\right) \tag{49}$$

in equation 48 with

$$\left(-\mu\frac{t}{T}\right)^{p} \tag{50}$$

where p takes a value other than 1·0.

7.6.3. *Sea wall slope*

Herbich *et al.* (1984) report results of 2-dimensional flume tests in which the angle of the sea wall with respect to the horizontal was varied as follows: 90° (vertical), 67·5°, 40°, 30°, 15°. A medium to coarse sized sand (d_{50} = 0·48 mm) was used in the tests, the wave period was in the range 1 to 2 seconds, and the water depth was around 0·13 to 0·22 m. The average (as opposed to the maximum) scour depth over a distance of 4·5 m in front of the wall was measured. Whilst this measure of the scour will mask the maximum values of the toe scour at the wall, the data followed a time development relationship similar to Equation (49) above. Herbich *et al.* comment that the results from their small-scale tests are at best qualitative. However, it is noted that in all cases the ultimate scour depths obtained were smaller than the incident wave height H, and larger for the same combination of wave condition and water depth in front of the 30° wall than in front of the 15° wall. The results for the scour depth at the 45°, 67·5° and 90° sea walls were approximately similar. In summary, the scour depth is only significantly reduced for sea wall slopes of 15° (1V:4H) or less.

7.6.4. *Angle of incidence*

The superimposed influence of oblique incidence waves and the reflected wave field from the sea wall leads to the formation of complex 3-dimensional short crested seas, high bottom velocities and increased vortex action. Hsu and Silvester (1989) and Silvester and Hsu (1997) suggested that this was the explanation for many scour problems in front of sea walls as the resulting wave field greatly increases the sediment suspension (Oumeraci, 1994). The work of O'Donoghue and Goldsworthy (1995) on normally incident waves is currently being extended by experiment and theory to the case of oblique incidence waves. This work will provide direct evidence for assessing the role of oblique waves on toe scour. Oblique waves will also generate a wave-induced current running along the face of the wall which can exacerbate the scour development (cf Section 7.7.1).

7.7. BREAKWATERS

Shore-attached breakwaters are commonly used to afford protection to harbour and dock facilities or the approach channels to ports. Essentially they are free-standing sea walls of various construction types. At many European and worldwide coastal locations detached breakwaters are used as shoreline protection structures.

7.7.1. *Long breakwater arms*

Depending upon the orientation of the breakwater to the prevailing wave direction, the scour development in front of the breakwater may or may not be similar to the sea wall case for normal incidence (2-dimensional). Scour development for wave angles normal to the alignment of the wall can be treated as 2-dimensional and akin to sea walls although, as with sea walls, even near normal incidence conditions may have a tendency to develop 3-dimensional bed topography.

For those situations in which the breakwater extends out from the shoreline the structure will become increasingly susceptible to the influence of tidal currents. For locations which have a large tidal current running orthogonal to the sea wall alignment the scour development at the tip of the breakwater will be similar to that found at the ends of spur dykes in rivers. Hence a similar approach (e.g. Hoffmans and Verheij, 1997) might be adopted to predicting the maximum scour depth, although the flow constriction may be less severe than in a river of finite width. Tidal current scour can also be significant along and at the ends of inlet training jetties and a compilation of data for the USA has been presented by Lillycrop and Hughes (1993).

Scour around breakwater heads for normal and near-normal incidence waves, including a co-directional current, can be predicted from the results of Gökçe *et al.* (1994). For the case of waves approaching perpendicular to the breakwater the depth of scour at the head of the breakwater, with KC based on the width of the breakwater head of order 1, is negligible (Figure 40, $L_p/D = 0$ for unprotected bed). For $KC \leq 1$ the flow is non-separated and for $KC > 12$ a horseshoe vortex is formed. The effect of bed protection in reducing scour depths is also shown in Figure 40 for two widths of scour protection. The action of a current past the breakwater head has been demonstrated to enhance significantly the scour depth. More extensive results for the vertical breakwater head and rubble breakwater head can be found in Sumer and Fredsøe (1997) and Fredsøe and Sumer (1997).

Funakoshi (1994) reports on the performance of breakwaters in Japan and comments that the construction of the breakwater triggers scour development in front of the breakwater typically up to a depth of around 2 m, in water depths of 12 to 18 m. The

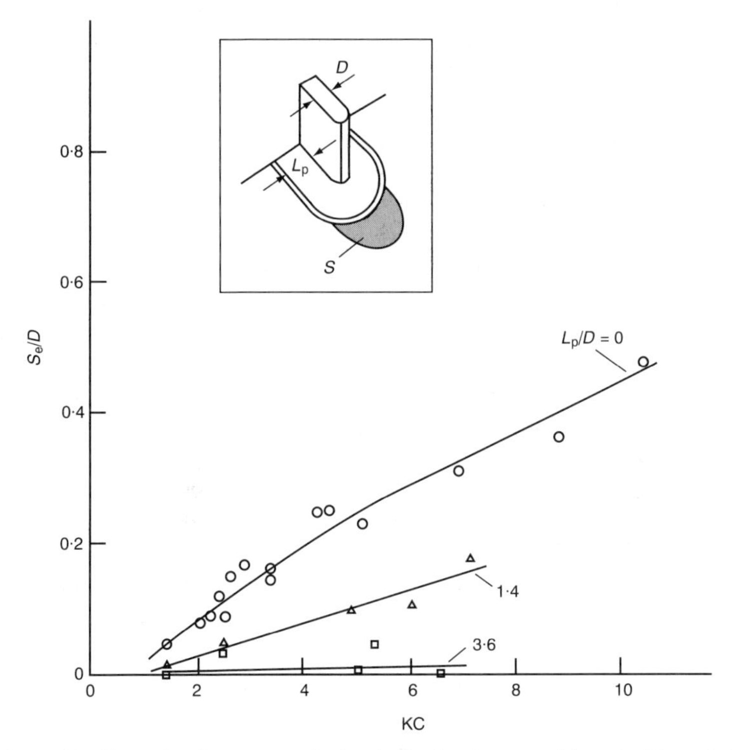

Figure 40. Scour development at the head of a breakwater with and without bed protection: live-bed conditions (reproduced from Gökçe et al., 1994, by permission of the Port and Harbour Research Institute, Yokohama)

maximum observed scour depths were up to 4 m and the scouring did not always occur at the breakwater heads. Depending upon the orientation of the breakwater he considered that the wave driven currents would exacerbate the scouring along the face of long breakwaters. Therefore each structure should be considered on a site by site basis as the orientation to the prevailing flows will be different. Further field evidence of scour and its consequences in various countries is discussed by Oumeraci (1994a) and Silvester and Hsu (1997).

7.7.2. Detached breakwaters

Detached breakwaters are used at many European and worldwide coastal locations as shoreline protection structures. The impact of the structures on the coastline has received considerable study in both the field and laboratory (e.g. Liberatore, 1992; Sawaragi, 1992) and numerically (Price *et al.*, 1995; O'Connor and Nicholson, 1995). The results obtained by Price *et al.* are shown in Figure 11. Indirectly some of these studies may provide information on local scour development but generally these results remain uncorrelated. Therefore, the full pattern of local scour which influences the performance and stability of the structure is not well documented.

Under certain circumstances, e.g. perpendicular wave attack, scour at the front face may be able to be predicted using a 2-dimensional approach for sea walls (e.g. Oumeraci, 1994b). Thus it is possible to produce scour profiles like those shown in Figure 38 along most of the length of the breakwater. However, because these structures are relatively short in the alongshore direction, the scour pattern can often be 3-dimensional. Scour around breakwater heads for normal incidence waves can be assessed from the results of Gökçe *et al.* (1994), see Figure 40. The action of a current flowing past the breakwater head in the same direction as the waves has been demonstrated to enhance significantly the scour depth.

The development of the full scour pattern due to normal and oblique incidence waves, including the littoral flow along the shoreline, has not been specifically monitored in the field or in controlled 3-dimensional laboratory tests. One aspect for further study is the transition zone from a 2-dimensional (e.g. Figure 38) to a 3-dimensional pattern at the breakwater head (e.g. Figure 40)—Powell and Whitehouse (1998).

7.8. FREE SETTLING OBJECTS

Structures or objects which are placed on the sea bed may undergo some settlement if the effective compressive strength of the soil underneath the bearing area is exceeded by the vertical pressure exerted across the footings. Settlement of heavy objects can occur in clay and this has been modelled for cylinders by Pastor *et al.* (1989). The same kind of self-weight settlement is not usually observed in sand and external forcing is required to initiate settlement. If the object is constrained from lateral movement then settling can occur through scouring of the soils which leads to a reduction in the bearing area under the object or through the cyclic pressures occurring under waves resulting in a loss of bearing strength (see Section 2.5.2). Other situations in which settlement operates include pipelines which can settle at span shoulders due to soil failure (see Section 7.4.16) and the interaction of spud cans with the sea bed (see Section 7.9). Nago and Maeno (1995) have demonstrated the settlement of a concrete block into sand under cyclic pressures. Some laboratory testing has been carried out on the scouring around submerged rectangular blocks with dimensions chosen to reflect generic types of subsea structure (Wagstaff, 1993). The stability of oil rig derived sea bed debris has been examined by Wimpey Offshore (1990).

If the objects lie unconstrained on the sea bed then they may subside into any scour pits that are formed. Carstens (1966) performed laboratory experiments to investigate the scouring and free-settling into the bed of short tubulars. The results showed that a steplike settling of the tubular into the sea bed could occur.

Unpublished work by the author (1993) has examined further the settlement of a free settling short cylinder (length:diameter = 5:1) due to scour. Scour takes place preferentially at the ends of the cylinder and the two scour pits extend towards the centre of the cylinder by tunnel erosion (similar to the scour under pipelines). The cylinder then settles into the bed by a lateral rocking motion as the tunnel erosion proceeds alternately in each of the two scour pits. Settlement is also characterised by a slow rolling motion of the cylinder into the scour pit formed on the upstream side. The settlement of the cylinder in steady flow can be predicted using the same kind of negative exponential time–development relationship already adopted for scour depth at piles and pipelines (Equation (3)) but with both a different shape to the curve and a modified form of the predictor for equilbrium scour depth. In this case the exponent p in Equation (3) determined from data is 0·6.

The equilibrium settlement depth predictor for steady flow has been determined from laboratory data as:

$$S_s = 0, \; 0 \leq U < 0{\cdot}75U_{cr} \tag{51a}$$

$$S_s = S_{smax} \frac{U - 0{\cdot}75U_{cr}}{0{\cdot}5U_{cr}}, \; 0{\cdot}75U_{cr} \leq U < 1{\cdot}25U_{cr} \tag{51b}$$

$$S_s = S_{smax}, \; 1{\cdot}25U_{cr} \leq U \tag{51c}$$

with

$$S_{smax} = 1{\cdot}15D \tag{51d}$$

In this expression the value $0{\cdot}75U_{cr}$ refers to the threshold for the onset of settlement as defined by the data as opposed to the threshold of scour.

Equations (51a–c) can be rewritten in terms of shear stress by assuming:

$$\frac{\tau}{\tau_{cr}} = \left(\frac{u}{u_{*cr}}\right)^2 = \left(\frac{U}{U_{cr}}\right)^2 \tag{52}$$

where u_*, u_{*cr} are the ambient and critical friction velocities. This is a reliable assumption so long as τ_{cr}, u_{*cr} and u_{cr} all relate to one sediment of constant roughness.

The time-scale for settlement in steady flow with $p = 0{\cdot}6$ is given by the equation in Figure 25, i.e. Equation (5a) with $A = 0{\cdot}095$ and $B = -2{\cdot}02$. Settlement in a tidally varying current can be predicted using the general approach in Section 2.2.4.

7.9. JACK-UP PLATFORMS

Jack-up rigs or barges are used in coastal and estuarine regions, as platforms for booster stations in the hydraulic transport of beach nourishment material (e.g. Angremond, 1992) in construction projects (e.g. for drilling sub-sea bed tunnels for outfalls or ballast handling), and for the installation of research platforms (De Wolf *et al.*, 1994) as well as for offshore oil and gas exploration. These platforms are floated to their location and then jacked-up to their operational level by the use of typically 3 or 4 mechanically operated legs. On the bottom of these legs are *spud cans* of varying shapes which provide the foundation support for the platform. Where these rigs are used in areas of soft soil the foundation performance needs to be continuously monitored, especially where the wave and current climate are competent enough to mobilise the sea bed sediments. Much of the research on the scour behaviour around spud cans is related to the requirements of the offshore oil and gas industry (e.g. Sweeney *et al.*, 1988). As a result of scour the jack-up leg can settle into the sea bed even during the normal short-term operation of a jack-up rig at a given location (Watson, 1979). The accurate determination of scour effects around jack-up legs and the subsequent settlement are important because of the limited length of the legs.

7.9.1. Field experience

Experiences of operating jack-up units in shallow water (13 m) with strong tidal currents offshore Canada (Song *et al.*, 1979) have highlighted the relative importance of sea bed scour and leg penetration. The problem encountered was that 2 to 3 m of scouring was expected and leg penetration of 4 to 5 m was expected which indicated bottom scour would not present a threat to the stability of the rig. However, in practice the penetration depth was only 2 m and the scour posed a treat to stability. To overcome this problem additional penetration was achieved using jetting/air lifting of sediment to place the caisson footing below the influence of scour.

Sweeney *et al.* (1988) measured the bearing area and scour pattern around a 3-legged jack-up offshore China. Their observations indicated that the scour pattern began to develop after the penetration of the three legs had stabilised and that eventually the scour holes around each leg coalesced to form a depression some 3 m deep and extending 15 m to 20 m from the outside rims of the spud cans (hexagonal shape, 12 m across). Sweeney *et al.* also found that the leg-penetration during storms was increased although, surprisingly, not in all cases.

Bijker and de Bruyn (1988) reported scour depths greater than 1 to $1{\cdot}5D$ for jack-ups in the presence of breaking waves along the Dutch coast.

7.9.2. Spud can–soil interaction

Once the spud can has been installed on the sea bed it settles until the soil resistance provided through the bearing area equals the weight of the platform. If the spud can still remains proud of the sea bed then the disturbance to the local flow field can result in scour and undermining of the footings. The spud can geometry and size are important in determining how much and where the sea bed scouring takes place. Where scouring of the soil reduces the bearing area to below a tolerable level the leg can be jacked down to a new position to increase the bearing area to an acceptable level. The scour-induced settlement process reaches an equilibrium when the spud can is completely buried in the soil. Therefore, because the leg and spud can settle into the sea bed under their own weight, or under additional loading of the leg, the scour/settlement of a jack-up leg is actually quite a complicated process.

Further settlement may take place independently of any scouring processes if the vertical load exceeds the bearing strength of the soil stratum on which the underside of the spud can lies. Therefore a detailed knowledge of the subsurface soil characteristics is essential. However, the case study reported by Song *et al.* (1979) has indicated that even if the scour can be predicted adequately the associated predictions of leg penetration may not be reliable in all cases.

7.9.3. Scour pattern

The scour pattern formed around the spud can in a steady current will take the form of a localised depression at the upstream side and an area of deposition at the downstream side. Thus, as well as reducing the total bearing area under the spud can scouring may produce an asymmetric bearing area. Direct measurements of the footprint shape and size have been made in steady flow laboratory tests by Sweeney *et al.* (1988). The footprint was monitored using conductance probes mounted flush with the underside of model spud can and the results reveal how the footprint characteristics vary with time and flow velocity (Figure 41).

7.9.4. Scour/settlement depth

Comprehensive data on the variations in the scour depth and associated leg settlement with flow velocity have not been published, although Sweeney *et al.* (1988) concluded from their tests that the scour-induced settlement could be increased from 0·6 m to 1·8 m with a doubling of the current velocity. The scour around semi-submersible drill rig footings has been studied experimentally by Wilson and Abel (1973) who predicted the occurrence of 6 m deep scour pits at positions around the three feet of the support structure. Tests on the scouring around spud can models have also been reported by Nagai *et al.* (1980). A scour/settlement model for the North Sea and any similarly moderate scour environment would need to be based on the interpretation of results from such laboratory experiments in conjunction with an orderly compilation of field data.

7.9.5. Spud can shape

Rapoport and Young (1988) have summarised the evolution of the spud can design for offshore applications since 1957, citing 25 different types, of various shapes and sizes.

Common sense suggests that the scouring around angular or tall spud cans is more vigorous, for the same soil and hydraulic conditions, than that around streamlined footings. Sweeney *et al.* (1988) reported the results from steady current flume tests with a triangular spud can which indicated that the scour-induced settlement was likely to be three times more severe than for the circular design. A comparison of the footprint development under the different shapes is shown in Figure 41. Lyons

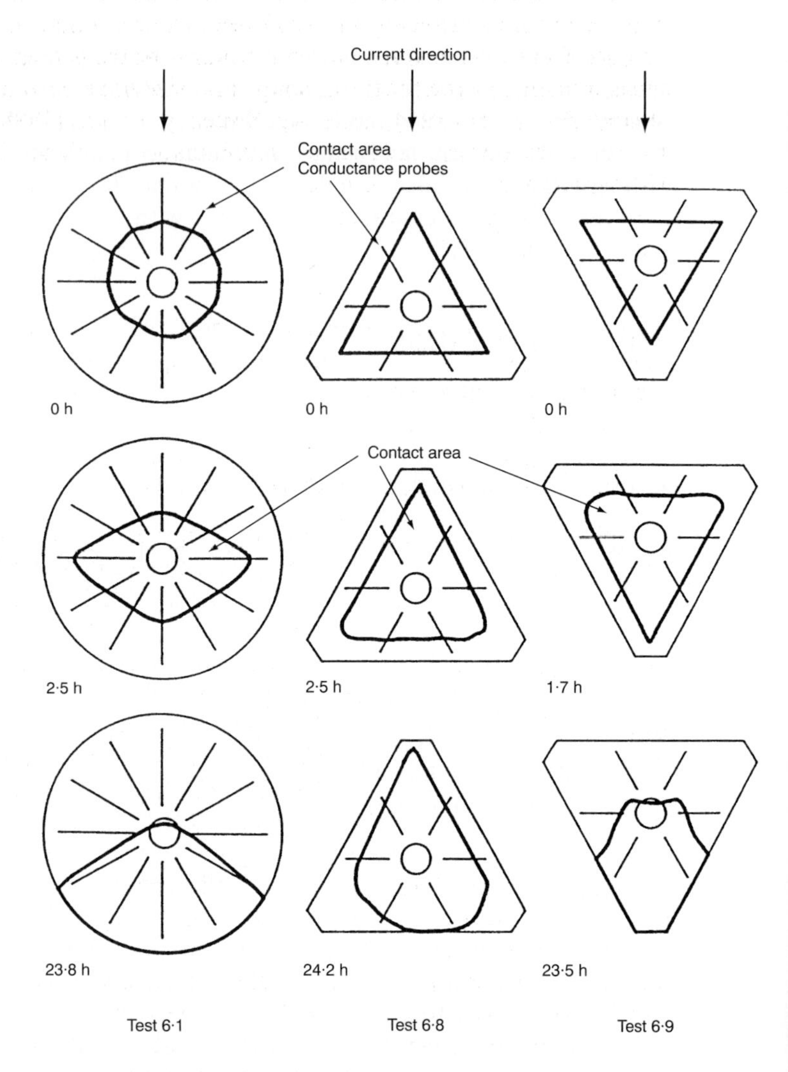

Figure 41. Footprint development measured under spud cans in laboratory tests (reproduced from Sweeney et al., 1988, by permission of the Offshore Technology Conference)

and Willson (1986) recommend the use of a relatively flat profile spud can in cases where scour was found to be a major problem, thus minimising the potential for scour. However, group effects leading to dishpan scour might still take place (Section 7.9.1).

Herbich *et al.* (1984) and Ninomiya *et al.* (1972) provide further data on the scour and settlement of jack-up legs of different shape.

7.9.6. Vertical load

From laboratory tests Sweeney *et al.* (1988) found that a doubling of the vertical load on the jack-up leg could halve the scour-induced settlement in sand.

7.9.7. Interactions with other structures

The proximity effects on the scour development around jack-up footings and jacket legs are important because of the effect of scour around the spud can on the existing piled foundations, especially where the installations may be separated by only a few metres (Lyons and Willson, 1986). Rapoport and Young (1988) reported the details of field investigations aimed at reducing the soil-foundation risks for mobile jack-up units. From a case history on underconsolidated clay the significant influence of the jack-up leg scour on settlement morphology of the sea bed became apparent (Figure 42*a*). The depth of the settlement pits left after departure of the jack-up rig was around 3·5 m for this location and their diameter was typically 2 spud can diameters (i.e. 24 m). The penetration of the jack-up legs to 70 ft also contributed to a significant reduction in the undrained shear strength of the soil (Figure 42*b*). Both of these effects could be significant for the fixity of piled structures or the stability of jack-up rigs operating in the vicinity of such pockmarks. Rapoport and Young suggest a 'rule-of-thumb' to avoid spud cans sliding into existing pockmarks—the edge to edge distance between the placed spud can and original spud can location should be at least one half of the spud can diameter.

The local scour pit formed around an existing jacket leg could influence the stability of mobile jack-ups when positioned alongside. Model tests were completed (Sweeney *et al.*, 1988)

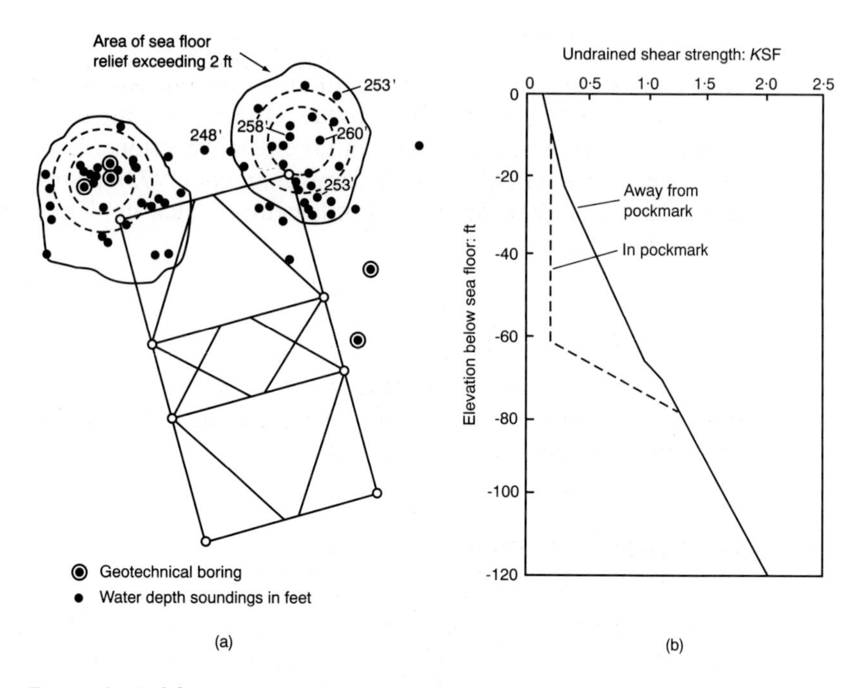

Figure 42. Bed features (a) and soil strength profile (b) left on the sea floor after jack-up removal (KSF = Kips per square foot) (reproduced from Rapoport and Young, 1988, by permission of Chapman and Hall)

for spud can jacket separation distances equal to one spud can diameter and one-half of the spud can diameter (see Figure 43). The results indicated that whilst the proximity of the jack-up rig did not significantly affect the scouring of the sand from around the jacket leg the scour-induced drop of the jack-up was consistently greater than in the case of the jack-up alone. In addition the settlement of the jack-up leg was increased in the closest proximity test. In these proximity tests the shape of the footprint under the spud can became extremely asymmetric and was observed to shift completely to one side of the soil spike at the centre of the spud can. Where the spud can overhangs the jacket scour pit (Figure 43) this could result in a sudden sliding of the jack-up leg into the scour pit.

It is clear that many interacting factors that have to be considered relating to the soil parameters and spud can position with respect to the sea bed when predicting the effect of storms

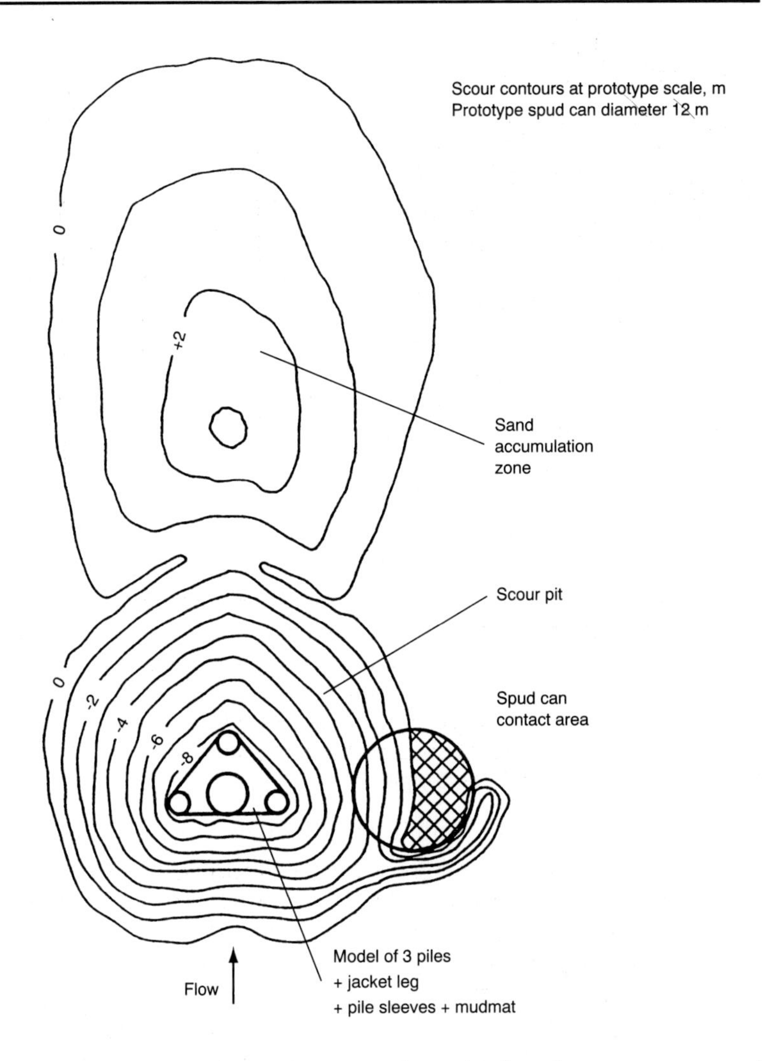

Figure 43. Scour interaction between jacket and jack-up legs on an unprotected sand bed (reproduced from Sweeney et al., 1988, by permission of the Offshore Technology Conference)

on scour-induced settlement. The scour and settling of jack-up spud cans should be analysed in the light of case history information where available and the appropriate laboratory data (Sweeney *et al.*, 1988). A further comparison of field observations with laboratory data is required to produce a comprehensive predictive model.

7.10. MISCELLANEOUS

7.10.1. *Scour due to jets (vertical, horizontal)*

An initial assessment of the scouring potential of submerged, vertical (axisymmetric) jets of water can be made from the results of Beltaos and Rajaratnam (1974). The wall shear stress in the impingement zone of the jet, acting on a smooth planar bed surface, can be predicted from

$$\tau_{jet} = 0{\cdot}16\rho \frac{U_o^2}{(z_n/d_o)^2} \tag{53}$$

where τ_{jet} is the maximum value of the bed shear stress, occurring at a radial distance r from the jet axis of $0{\cdot}14r/z_n$, z_n is the height of the nozzle (diameter d_o) above the bed, ρ the density of the jetting and ambient water and U_o the jet velocity at the nozzle (discharge Q divided by nozzle area). The scour potential can thus be assessed from the ratio τ_{jet}/τ_{cr} for the sediment and the depth and radial extent of scour in cohesionless sediments can be estimated using empirical relationships (e.g. Rajaratnam, 1982) involving the densimetric Froude Number,

$$F_o = \frac{U_o}{[(s-1)\,gd_{50}]^{\frac{1}{2}}} \tag{54}$$

Recently, Chiew and Lim (1996) have extended this type of analysis to predict the depth, length and width of scour pits formed by horizontal jets, those issuing both at the bed and at a distance above the bed. Scour caused by jet and culvert flows is discussed further by Hoffmans and Verheij (1997).

The response of cohesive sediments to jets does not appear to have received much attention (Breusers and Raudkivi, 1991) and the scour pit characteristics and scour rate will be controlled by the sub-surface bed properties. The results of unpublished tests performed by the author at HR Wallingford, for a vertical jet eroding a cohesive sediment bed, suggest a close relationship between the erosion flux from the bed and the total kinetic energy issuing from the nozzle ($\sim QU_o^2$).

7.10.2. *Scour due to manoeuvring vessels*

During the construction of berthing and harbour works it will be necessary to evaluate the potential for scouring of the bed sediments due to the flow induced by the passage of the vessel and the action of the propeller wash (e.g. Hamill, 1980; Verheij, 1983; CIRA/CUR, 1991). Hamill and Verheij both provide expressions for the scour depth caused by propeller wash as a function of F_0, propeller characteristics and height of the propeller above the bed. The likely effects on the bed material of the flow induced by the moving vessel and the propeller wash can be assessed directly from published data or by physical model tests. As a result bed protection can be designed to withstand the scouring forces.

7.10.3. *Scour under stationary vessels*

The potential for scour damage due to flow under and around a stationary vessel (or moored structure) in shallow water can be assessed from a calculation of the flow field using potential flow theory (e.g. McDougal and Sulisz, 1989). By combining the flow model with a (non-coupled) soil model (Biot) the model results allow the potential for liquefaction and scour to be estimated from the incident wave conditions and the reflection coefficient of the structure. A more comprehensive investigation of the bed stability and subsequent morphological development can be obtained in a physical model study.

7.10.4. *Wrecks*

The sea bed deformation observed around wrecks (or other obstacles such as boulders) on the sea bed can provide a good indication of whether the bed sediments are mobile and of the prevailing sediment transport direction.

The scouring and settlement behaviour of wrecked vessels and vessels grounded on, for example, submerged sand banks can be determined from the analysis of available case histories or by physical model tests.

7.11. SUMMARY

This chapter has summarised available knowledge on scour prediction methods or typical scour development by currents or waves. Where available, methods to predict (or values for) the following have been given:

(a) maximum scour depth in
- current flow
- wave flow
- wave plus current flow

(b) time development of scour in
- current flow
- wave flow
- wave plus current flow.

For most cases there is no generally accepted method for predicting the scour due to combined wave and current flows, or under the action of a time-varying wave–current magnitude and direction. In addition the influence of breaking waves on scour development (e.g. at piled structures) has not been studied in more than a few individual cases. The available results indicate the importance of breaking waves in exacerbating scour and the need for further research. Another area requiring further research effort is in the scour of cohesive sediments which could be especially important for structures in estuaries. In all scour predictions the independent long-term variations in the general bed level and the variability in the bed level due to bedforms need to be considered. The scour at complex installations for which case histories are not applicable should be investigated through physical model tests or hybrid physical–numerical modelling studies, and ultimately these predictions should be validated at prototype scale.

References

Abou-Seida, M. M. (1963). *Sediment scour at structures*, Technical Report HEL-4-2, University of California, Berkeley, California.

Angremond, K. d' (1992). Beach nourishment, in: *Design and Reliability of Coastal Structures*, Proceedings of the Short Course attached to 23rd ICCE, Venice, October 1992, pp. 415–432.

Angus, N. M. and Moore, R. L. (1982). Scour repair methods in the Southern North Sea. Paper OTC 4410 in *Proceedings 14th OTC Conference*, 1982, pp. 385–389 plus 10 pages tables and figures.

Armbrust, S. F. (1982). *Scour about a cylindrical pile due to steady and oscillatory motion*. MSc thesis, Texas A & M University.

Baker, C. J. (1979). The laminar horseshoe vortex. *J. Fluid. Mech.*, **95**, 347–367.

Baker, C. J. (1980). The turbulent horseshoe vortex. *J Wind Eng. Ind. Aerodyn.*, 9–23.

Basak, V., Basamisli, Y. and Ergun, O. (1975). Maximum equilibrium scour depth around linear-axis square cross-section pier groups, *Devlet su isteri genel mudurlugi*, Report 583, Ankara (in Turkish).

Bayazit, M. (1967). Resistance to reversing flows over movable beds. *J. Hydraul. Div. Am. Soc. Civ. Engrs*, **95**, (HY4), 1109–1127.

Bělik, L. (1973). The secondary flow about circular cylinders mounted normal to a flat plate. *Aeronaut. Q.*, Feb., 47–54.

Beltaos, S. and Rajaratnam, N. (1974). Impinging circular turbulent jets. *J. Hydraul. Div. Am. Soc. Civ. Engrs*, **100**, (HY10).

Bettess, R. (1990). Survey of lightweight sediments for use in mobile-bed physical models, in: *Movable Bed Physical Models*, pp. 115–123, Kluwer Academic Publishers, Dordrecht.

BGS (1987). *Sea bed sediments around the United Kingdom*. 1:1 000 000 map (North sheet and South sheet). (Also at 1:250 000 scale for selected areas.) British Geological Survey, Natural Environment Research Council.

Bijker, E. W. (1986). Scour around structures, *Proc. 21st Int. Conf. Coastal. Engng.*, 1754–1768, ASCE.

Bijker, E. W. and de Bruyn, C. A. (1988). Erosion around a pile due to current and breaking waves, *Proc. 21st Int. Conf. Coastal Engng.*, Vol. 2, Malaga.

Bijker, E. W. and Leeuwenstein, W. (1984). Interaction between pipelines and the sea bed under the influence of waves and currents, in: *Sea Bed Mechanics*, Proc. IUTAM-IUGG Symp, ed. B. Denness, Graham and Trotman, London, 235–242, September.

Bishop, J. R. (1980). Experience with scour at the Christchurch Bay tower, *Proc. one day Society for Underwater Technology seminar on scour prevention techniques around offshore structures*, 11–22, 16 December.

BODC (1991). *United Kingdom Digital Marine Atlas*, British Oceanographic Data Centre, Proudman Oceanographic Laboratory, Bidston, January 1991 version.

Breusers, H. N. C. (1972). *Local scour near offshore structures*, Delft Hydraulics Publication 105, Delft.

Breusers, H. N. C. (1975). *Computation of velocity profiles in scour holes*, Delft Hydraulics Laboratory, Publication 152.

Breusers, H. N. C., Nicollet, G. and Shen, H. W. (1977). Local scour around cylindrical piers, *J. Hydr. Res.*, **15**, (3), 211–252.

Breusers, H. N. C. and Raudkivi, A. J. (1991). *Scouring*, IAHR-AIRH Hydraulic Structures Design Manual, 2, Balkema, Rotterdam.

Brørs, B. (1993). *Local scour around pipelines and piers: a review of numerical modelling*. Unpublished.

Brørs, B. (1997). Numerical modelling of flow and scour at pipelines. *J. Hydraul. Engng, Am. Soc. Civ. Engrs*, submitted.

Carpenter, K. and Powell, K. A. (1998). *Toe scour at vertical seawalls. Mechanisms and prediction methods.* HR Wallingford Report SR 506.

Carstens, M. R. (1966). Similarity laws for localised scour, *Proc. Am. Soc. Civ. Engrs, J. Hydraul. Div.*, **92**, (3), 13–36.

CDIT (1986). *Recent Port and Harbor Engineering Technology in Japan*, Coastal Development Institute of Technology, Japan.

Chabert, J. and Engeldinger, P. (1956). *Etude des affouillements autour des piles de ponts*, Laboratoire National d'Hydralique, Chatou, Paris.

Chesher, T. J., Wallace, H. M., Meadowcroft, I. C. and Southgate, H. N. (1993). *A Morphodynamic coastal area model first annual report.* HR Wallingford Report SR 337.

Chiew, Y. M. (1991a). Flow around horizontal circular cylinder in shallow flows, *Proc. Am. Soc. Civ. Engrs, J. Waterway, Port, Coastal, and Ocean Engng*, **117**, (2), 120–135.

Chiew, Y. M. (1991b). Prediction of maximum scour depth at submarine pipelines, *Proc. Am. Soc. Civ. Engrs, J. Hydraulic Engng*, **117**, (4), 452–466.

Chiew, Y. M. (1993). Effect of spoilers on wave-induced scour at submarine pipelines, *Proc. Am. Soc. Civ. Engrs, J. Waterway, Port, Coastal, and Ocean Engng*, **119**, (4), 417–429.

Chiew, Y. M. and Lim, S. Y. (1996). Local scour by a deeply submerged horizontal circular jet, *J. Hydraul. Engn.*, **122**, (9), 529–532.

Chow, W.-Y. and Herbich, J. B. (1978). Scour around a group of piles, Paper OTC 3308, *Offshore Technology Conference*, Houston, 1978.

CIRIA (1986). *Sea walls: survey of performance and design practice*, CIRIA technical note 125.

CIRIA (1996). *Beach management manual*, CIRIA report 153.

CIRIA/CUR (1991). *Manual on the use of rock in coastal and shoreline engineering*. CIRIA special publication 83/CUR report 154.

Clark, A. and Novak, P. (1984). Local erosion at vertical piles by waves and currents, in: *Sea Bed Mechanics*, Proc. IUTAM-IUGG Symp, ed. B. Denness, Graham and Trotman, London, September, 243–249.

Clark A., Novak, P. and Russell, K. (1982). Modelling of local scour with particular reference to offshore structures, *Proc. BHRA Int. Conf. Hydraulic Modelling of Civil Engineering Structures*, Coventry, 411–422.

Dahlberg, R. (1981). Observation of scour around offshore structures, *Symp. Geotech. Aspects of Coastal and Offshore Structures*, Bangkok, 159–172, December.

Davies, A. G., Soulsby, R. L. and King, H. L. (1988). A numerical model of the combined wave and current bottom boundary layer. *J. Geophys. Res.*, **93**, 491–508.

De Vriend, H. J., Zyserman, J., Nicholson, J., Roelvink, J. A., Péchon, P. and Southgate, H. N. (1993). Medium term 2DH coastal area modelling, *Coastal Engng*, **21**, 193–224.

De Wolf, P., Van Den Bergh, P., Thomas, R., Lanckzweirt, M. and Hyde, P. (1994). Measuring platforms in the North Sea in Belgian territorial waters and on the Belgian continental shelf, *PIANC Bulletin*, **82**, 38–51.

Delo, E. A and Ockenden, M. C. (1992). *Estuarine muds manual.* HR Wallingford Report SR309.

Department of Energy/Health and Safety Executive (1992/93). *Offshore Installations: guidance on design, construction and certification.* HMSO, London, 4th Edition (revised).

Di Natale, M. (1991). Scour around cylindrical piles due to wave motion in the surf-zone, *Coastal Zone '91*, Long Beach, California, July.

Diamantidis, D. and Arnesen, K. (1986). Scour effects in piled structures—a sensitivity analysis, *Ocean Engng*, **13**, (5), 497–502.

Dietz, J. W. (1995). Current pattern, scouring and bottom protection at the Eider barrier. *Mitteilungsblatt der Bundesanstalt für Wasserbau*, **73**, (27–109) (in German).

Draper, L. (1991). *Wave Climate Atlas of the British Isles*, Department of Energy, Offshore Technology Report, OTH 89 303, HMSO, London.

Dyer, K. R. (1986). *Coastal and Estuarine Sediment Dynamics*. John Wiley and Sons, Chichester.

Eckman, J. E. and Nowell, A. R. M. (1984). Boundary skin friction and sediment transport about an animal mimic tube, *Sedimentology*, **31**, 851–862.

Escarameia, M. and May, R. W. P. (1995). Stability of riprap and concrete blocks in highly turbulent flows. *Proc. Instn Civ. Engrs Wat. Marit. Energy*, **112**, 227–237.

Ettema, R. (1990). Design method for local scour at bridge piers, *Proc. Am. Soc. Civ. Engrs, J. Hydraulic Engng*, **116**, (10), 1290–1292.

Fowler, J. E. (1993). *Coastal scour problems and methods for prediction of maximum scour*. Coastal Engineering Research Center, US Army Corps of Engineers, Technical Report CERT-93-8.

Franzetti, S., Larcan, E. and Mignosa, P. (1981). *Erosione localizzata alla base delle pile dei ponti: considerazioni sui risultati di unindagine sperimentale su modello*, Istituto di Idraulica e Construzioni Idrauliche del Politecnico di Milano, N.296 (in Italian).

Franzetti, S., Larcan, E. and Mignosa, P. (1982). Influence of tests duration on the evaluation of ultimate scour around circular piers, *Proc. BHRA Int. Conf. Hydraulic Modelling of Civil Engineering Structures*, Coventry, 381–396.

Fredsøe, J. (1978). *Experiments of natural backfilling of pipeline trenches*, Inst. Hydrodynamics and Hydraulic Engng, Tech. University Denmark, Prog. Rept 46, 3–6.

Fredsøe, J. and Sumer, B. M. (1997). Scour at the round head of a rubble-mound breakwater, *Coastal Engng*, **29**, 231–262.

Fredsøe, J., Sumer, B. M. and Arnskov, M. M. (1992). Time scale for wave/current scour below pipelines, *Int. J. Offshore Polar Engng.*, **2**, (1), 13–17.

Fredsøe, J., Hansen, E. A., Mao, Y. and Sumer, B. M. (1988). Three-dimensional scour below pipelines. *J. Offshore Mech. Arctic Engng*, **110**, 373–379.

Funakoshi, H. (1994). Survey of long-term deformation of composite breakwaters along the Japan Sea. *Proc. Int. Workshop on Wave Barriers in Deep Waters*, PHRI Yokosuka, 10–14 January, 239–266.

Gökçe, K. T. and Gunbak, A. R. (1992). Effect of spoilers on self burial of submarine pipelines by waves, Abstract 293, *23rd Int. Conf. Coastal Engng*, October.

Gökçe, T., Sumer, B. M. and Fredsøe, J. (1994). Scour around the head of a. vertical-wall breakwater. *Proceedings of HYDRO-PORT 94*, 19–21 October, Yokosuka, Port and Harbour Research Institute, 1097–1116.

Grace, R. A. (1980). *Marine Outfall Systems: Planning, Design and Construction*. Prentice Hall, Hemel Hempstead.

Grass, A. J. and Hosseinzadeh-Dalir, A. (1995). A simple theoretical model for estimating maximum scour depth under sea bed pipelines. *Proc HYDRA 2000*, Vol. 3, XXVIth IAHR Congress, London, Thomas Telford, 281–287.

Hales, L. Z. (1980a). *Erosion control of scour during construction*. Report 1: Present design and construction practice. Hydraulics Laboratory, US Army Engineer Waterways Experiment Station, Vicksburg, Technical Report (Series) HL-80-3.

Hales, L. Z. (1980b). *Erosion control of scour during construction* Report 2: Literature survey of theoretical, experimental and prototype investigations. Hydraulics Laboratory, US Army Engineer Waterways Experiment Station, Vicksburg, Technical Report (Series) HL-80-3.

Hales, L. Z. and Houston, J. R. (1983). *Erosion control of scour during construction*; Report 4: stability of underlayer material placed in advance of construction to prevent scour; hydraulic model investigation. Hydraulics Laboratory, US Army Engineer Waterways Experiment Station, Vicksburg, Technical Report (Series) HL-80-3.

Hamill, G. A. (1988). Scouring action of the propeller jet produced by a slowly moving ship, *PIANC Bulletin*, 62, 85–110.

Hannah, C. R. (1978). *Scour at pile groups*. University of Canterbury, NZ, Civil Engng, Report 216.

Hansen, E. A., Fredsøe, J. and Ye, M. (1986). Two-dimensional scour below pipelines, *Proc. 5th Int. Offshore Mechanics and Arctic Engng (OMAE) Symp.*, Vol. 3, 670–678.

Harford, C. and Ramsay, D. (1996). *A catalogue of synthetic wave data from around the coast of England and Wales*. HR Wallingford Report SR373.

Harley, M. (1992). *Geotextiles for scour rectification*. Unpublished MSc thesis, Cranfield Institute of Technology.

Hebsgaard, M., Ennemark, F., Spangenberg, S., Fredsøe, J. and Gravesen, H. (1994). Scour model tests with bridge piers; *PIANC Bulletin*, **82**, 84–92.

Heibaum, M. (1995). Scour stabilisation at the Eider storm surge barrier—geotechnical stability of revetment and subsoil. *Mitteilungsblatt der Bundesanstalt für Wasserbau*, **73**, (111–122) (in German).

Herbich, J. B. (1985). Hydromechanics of submarine pipelines: design problems, *Can. J. Civ. Engng*, **12**, 863–874.

Herbich, J. B., Schiller, R. E., Watanabe, R. K., Dunlop, W. A. (1984). *Seafloor Scour: Design Guidelines For Ocean-founded Structures*, Marcel Dekker, New York and Basel.

Hicks, D. M. and Green, M. O. (1997). The 'fall-speed' parameter as an index of cross-shore sand transport: verification from measurements at the shoreface. *Proc. Pacific Coasts and Ports '97 Conf.*, Christchurch, University of Canterbury, New Zealand, 1089–1094.

Hindmarsh, F. R. (1980). Experiences with artificial seaweed for scour prevention. Proceedings of a one-day seminar, *Scour Prevention Techniques around Offshore Structures*, Society for Underwater Technology, London, 27–34.

Hirai, S. and Kurata, K. (1982). Scour around multiple- and submerged circular cylinders, *Memoirs Faculty of Engineering*, Osaka City Univ., **23**, 183–190.

Hjorth, P. (1975). Studies on the nature of local scour. *Bull. Ser. A*, 46, Dept Of Water Resources Engng, Lund Inst. Tech., Lund.

Hoffmans, G. J. M. C. and Booij, R. (1993). Two-dimensional modelling of local-scour holes. *J. Hydraul. Res. IAHR*, **31**, (5), 615–634.

Hoffmans, G. J. M. C. and Verheij, H. J. (1997). *Scour Manual*. Balkema, Rotterdam.

HR Wallingford (1972). *Field tests on the behaviour of pipes when laid on the sea bed and subjected to tidal currents*. HR Wallingford Report INT 113 (internal).

HR Wallingford (1993). *South coast sea bed mobility study: summary report*. HR Wallingford Report EX 2795.

Hsu, J. R. C. and Silvester, R. (1989). Model test results of scour along breakwater. *J. Wat. Port. Coast. Ocean Engng, Am. Soc. Civ. Engrs*, **115**, 1, 66–85.

Hughes, S. A. (1993). *Physical Models and Laboratory Techniques in Coastal Engineering*, World Scientific.

Hulsbergen, C. H. (1986). Spoilers for stimulated self-burial of submarine pipelines, Paper OTC 5339, *Offshore Technology Conference*, Houston, 441–443, May.

ICE (1985). *Coastal Engineering Research*. Report prepared on behalf of the Institution of Civil Engineers Maritime Engineering Group, Thomas Telford, London.

Imberger, J., Alach, D. and Schepis, J. (1982). Scour around circular cylinder in deep water, *Proc. 18th Int. Conf. Coastal Engng*, Vol. 2, 1522–1554, ASCE.

Ingram, R. (1993). 3D Sonargraphics: a sea bed survey tool, Acoustic classification and mapping of the sea bed, *Proc. Inst. Acoustics*, **15**, (2), 335–342.

Irie, I. and Nadaoka, K. (1984). Laboratory reproduction of sea bed scour in front of breakwaters. *Proc. 19th Int. Conf. Coastal Engng*, 1715–1731, ASCE.

Jinsi, B. (1986). Additional stabilization of submarine pipelines. *Proc. 5th International Offshore Mechanics and Arctic Engineering Symposium*, Tokyo, Vol. 3.

Jonsson, I. G. and Carlsen, N. A. (1976). Experimental and theoretical investigations in an oscillatory turbulent boundary layer. *J. Hydraul. Res.*, **14**, 45–60.

Katsui, H. and Bijker, E. C. (1986). Expected transport rate of material on sea bed. *J. Hydraul. Engng, Am. Soc. Civ. Engrs*, **112**, (9), 861–867.

Katsui, H. and Toue, T. (1992). Bottom shear stress in coexistent field of superimposed waves and current and scouring around a large-scale circular cylinder, *Coastal Engng in Japan*, **35**, (1), 93–110.

Kawata, Y. and Tsuchiya, Y. (1988). Local scour around cylindrical piles due to waves and currents combined, *Proc. 21st Int. Conf. Coastal Engng*, *2*, 1310–1322, Malaga.

Kjeldsen, S. P., Gjørsvik, O. and Bringaker, K. G. (1974). *Experiments with local scour around submarine pipelines in a uniform current*, River and Harbour Laboratory, Technical University of Norway, Trondheim, VHL/SINTEF Report STF60 A73085.

Knott, T. (1988). Article in *Offshore Engineer*, May.

Kothyari, U. C., Garde, R. J. and Ranga Raju, K. G. (1992). Temporal variation of scour around circular bridge piers, *Proc. Am. Soc. Civ. Engrs, J. Hydraul. Engng*, **118**, (8), 1091–1106.

Kraus, N. C. and McDougal, W. G. (1996). The effects of seawalls on the beach: Part 1, an updated literature review. *J. Coastal Res.*, **12**, (3), 691–701.

Kroezen, M., Vellinga, P., Lindenberg, J. and Burger, A. M. (1982). *Geotechnical and hydraulic aspects with regard to sea bed and slope stability*, Delft Hydraulics Laboratory, Publication 272, Delft.

Langhorne, D. N. (1980). *A preliminary study of the Haisborough Sand sandwave field and its relevance to submarine pipelines*. Institute of Oceanographic Sciences Report 100.

Lee, K. L. and Focht, J. A. (1975). Liquefaction potential at Ekofisk tank in North Sea, *Proc. J. Hydraul. Engng, Am. Soc. Civ. Engrs, J. Geotech. Engng Div.*, **101**, (GT1), 1–18.

Leeuwenstein, W., Bijker, E. W., Peerbolte, E. B. and Wind, H. G. (1985). The natural self-burial of submarine pipelines, *Proceedings BOSS*, 717–728.

Leeuwenstein, W. and Wind, H. G. (1984). The computation of bed shear in a. numerical model. *Inernational Coastal Engineering Conference 1984 Proceedings*, 1685–1702 (and Publication 389, Delft Hydraulics Laboratory (1985).

Liberatore, G. (1992): Detached breakwaters and their use in Italy,. in: *Design and Reliability of Coastal Structures*, Proceedings of the Short Course attached to 23rd ICCE, Venice, October, 373–395.

Lillycrop, W. J. and Hughes, S. A. (1993). *Scour hole problems experienced by the Corps of Engineers; data presentation and summary*. US Army Corps of Engineers, Waterways Experiment Station, PO Box 631, Vicksburg, Miss. 39180. Miscellaneous paper. CERC-93-2.

Loveless, J. H. and Grant, G. T. (1995). Physical modelling of scour at coastal structures. *Proc. HYDRA 2000*, Vol. 3, XXVIth IAHR Congress, London, Thomas Telford, 293–298.

Lucassen, R. J. (1984). *Scour underneath submarine pipelines*. MSc thesis, Department of Civil Engineering, Delft University of Technology.

Lyons, R. H. and Willson, S. (1986). Effects of spud cans on adjacent piled foundations, in: *The Jack-Up Drilling Platform: Design and Operation*, ed. L. F. Boswell, Collins, Chapter 2.

Machemel, J. L. and Abad, G. (1975). Scour around marine foundations, Paper OTC 2313, *Offshore Technology Conference*, Dallas.

Maidl, B. and Schiller, W. (1979). Testing experiences of different scour protection technologies in the North Sea. Paper OTC 3470, *Proc. 11th Offshore Technology Conference*, Houston, 981–987.

Maidl, B. and Stein, D. (1981), New experiences in scour protection for offshore platforms and pipelines, *Symp. Geotech. Aspects of Coastal and Offshore Structures*, Bangkok, 173–186, December .

Mann, K. (1991). *Sedimentation at jetties*, HR Wallingford Report SR 285.

Mao, Y. (1986). *The interaction between a pipeline and erodible bed*, Inst. Hydrodynamics and Hydraulic Engng, Tech. University Denmark, Series paper 39, March.

May, R. P. and Willoughby, I. R. (1990). *Local scour around large obstructions*, HR Wallingford Report SR 240.

McCarron, W. O. and Broussard, M. D. (1992). Measured jack-up response and spudcan–seafloor interaction for an extreme storm event, *Proceedings BOSS*.

McDougal, W. G., Kraus, N. C. and Ajiwibowo, H. (1996). The effects of seawalls on the beach: Part II, numerical modelling of SUPERTANK seawall tests. *J. Coastal Res.*, **12**, (3), 702–713.

McDougal, W. G. and Sulisz, W. (1989). Sea bed stability near floating structures. *J. Waterway, Port, Coastal and Ocean Engng*, **115**, (6), 727–739.

Mitchener, H. J., Torfs, H. and Whitehouse, R. J. S. (1996). Erosion of mud/sand mixtures. *Coastal Engng*, **29**, 1–25. [Erratum, **30** (1997) 319].

Müller, G. (1995). Wave impact pressures on a vertical wall and their effect on sea bed pressures. *Proceedings of HYDRA 2000*, Vol. 3, XXVIth IAHR Congress, London, Thomas Telford, 275–280.

Myrhaug D. (1995). Bottom friction beneath random waves. *Coastal Engng*, **24**, 259–273.

Myrhaug, D. and Slaattelid, O. H. (1989). Combined wave and current boundary layer model for fixed rough sea beds, *Ocean Engng*, **16**, 119–142.

Nagai, S., Oda, K. and Kurata, K. (1980). *Scour around leg-spudcans of drilling rigs*, Osaka City University, Osaka.

Nago, H. and Maeno, S. (1995). Settlement of concrete block into sand bed under cyclic loading of water pressure. *Proceedings of HYDRA 2000*, XXVIth IAHR Congress, London, Thomas Telford, 317–321.

Niedoroda, A. W. and Dalton, C. (1982).A review of the fluid mechanics of ocean scour. *Ocean Engng*, 9, (2), 59–170.

Ninomiya, K., Tagaya, K. and Murase, Y. (1972). A study on suction and scouring of sit-on-bottom type offshore structure, Paper OTC 1605, *Offshore Technology Conference*, Dallas, May.

Noble Denton (1984). *Environmental Parameters on the United Kingdom Continental Shelf.* Report prepared by Noble Denton and Associates Ltd for Department of Energy—Offshore Technology Report OTH 84 201, HMSO, London.

Ockenden, M. C. and Soulsby, R. L. (1994). *Sediment transport by currents plus irregular waves.* HR Wallingford report SR 376.

O'Connor, B. A. and Clarke, C. S. J. (1986). The dishpan scour problem, *Proc. 3rd Indian Conf. Ocean Engng*, Bombay, Vol. 2, J1–J12.

O'Connor, B.A. and Nicholson, J. (1995). Applications of a coastal area morphodynamic model. Article 7–15 in *Advances in Coastal Morphodynamics: An overview of the G8-Coastal Morphodynamics Project*, eds Stive, M. J. F, De Vriend, H. J., Fredsøe, J., Hamm, L., Soulsby, R. L., Teisson, C. and Winterwerp, J. C., Delft Hydraulics, Delft.

O'Donoghue, T. and Goldsworthy, C. J. (1995). Near-bed velocities in front of seawalls. *Proceedings of HYDRA 2000*, Vol. 3, XXVIth IAHR Congress, London, Thomas Telford, 311–316.

O'Riordan, N. J. and Clare, D. G. (1990). Geotechnical considerations for the installation of gravity base structures, Paper OTC 6381, *22nd Offshore Technology Conference*, Houston, 309–316, May.

Olsen, N. R. B. (1991). *A three-dimensional numerical model for simulation of sediment movements in water intakes.* Dr Ing, thesis, University of Trondheim,

Olsen, N. R. B. and Melaaen, M. C. (1993). Three-dimensional calculation of scour around cylinders, *Proc. Am. Soc. Civ. Engrs, J. Hydraul. Engng*, **119**, (9), 1048–1054.

Oumeraci, H. (1994a). Review and analysis of vertical breakwater failures—lessons learned. *Coastal Engng*, **22**, 3–29.

Oumeraci, H. (1994b). Scour in front of vertical breakwaters – Review of problems. *Proceedings of International Workshop on Wave Barriers in Deep waters*, Port and Harbour Research Institute, Yokosuka, Japan, 281–307.

Pastor, J., Turgeman, S. and Avallet, C. (1989). Predicting the phenomenon of burying through gravity in purely cohesive sedimentary sea beds, *Géotechnique*, **39**, (4), 625–639.

Pluim-van der Velden, E. T. J. M. and Bijker, E. W. (1992). Local scour near submarine pipelines on a cohesive bottom, *Proceedings of the 6th International Conference on the Behaviour of Offshore Structures*, Supplement, Bentham Press.

Posey, C. J. (1970). Protection against underscour, Paper OTC 1304, *Offshore Technology Conference*, Dallas, April.

Posey, C. J. and Sybert, J. H. (1961). Erosion protection of production structures. *Proceedings 9th IAHR Congress*, 1157–1162.

Powell, K. A. (1987). *Toe scour at seawalls subject to wave action: a literature review*. HR Wallingford Report SR119.

Powell, K. A. and Lowe, J. P. (1994). The scouring of sediments at the toe of sea walls. *Proc. Hornafjordor Int. Coastal Symposium*, Iceland, 20–24 June, ed. G. Viggosson.

Powell, K. A. and Whitehouse, R. J. S. (1998). The occurrence and prediction of scour at coastal and estuarine structures. *Proc. MAFF Conf. River and Coastal Engrs*, University of Keele, 1–3 July, 1998.

Price D. M., Chesher T. J. and Southgate, H. N. (1995). Medium-term morphodynamic coastal area modelling. Article 7.7 in *Advances in Coastal Morphodynamics: An overview of the G8-Coastal Morphodynamics Project*, eds Stive, M. J. F, De Vriend, H. J., Fredsøe, J., Hamm, L., Soulsby, R. L., Teisson, C. and Winterwerp, J. C., Delft Hydraulics, Delft.

Puls, W. (1981). Numerical simulation of bedform mechanics. *Mitteilungen des Institut für Meereskunde der Universität Hamburg*, Hamburg, 1–147.

Rajaratnam, N. (1982). Erosion by submerged circular jets. *J. Hydraul. Div., Am. Soc. Civ. Engng*, **108**, (HY2).

Rance, P. J. (1980). The potential for scour around large objects, in: *Scour Prevention Techniques around Offshore Structures*, Proceedings of a one day seminar, 16 December 1980, Society for Underwater Technology, London. 41–53.

Rapoport, V. and Young, A. G. (1988). Foundation performance of jack-up drilling units, analysis of case histories, *Proc. Conf. Mobile Offshore Structures*, City University, London, September, Eds Boswell, L. S, D'Mello, C. A. and Edwards, A. K, Chapman and Hall, London.

Reddy, D. V., Arockiasamy, M. and Jani, J. S. (1990). Structural safety of an ocean outfall against hurricane damage. *Marine Struct.*, **3**, 25–41.

Reese, L. C., Wang, S. T. and Long, J. H. (1989). Scour from cyclic loading of piles, paper OTC 6005, *Proc. 21st Annual OTC*, Houston, 395–402.

Rodi, W. (1984). *Turbulence models and their application in hydraulics*. IAHR State-of-the-art Paper, Delft.

Roelofsen, N. (1980). Scour control using dredging technology. *Scour Prevention Techniques around Offshore Structures*, Proceedings of a one-day seminar, 16 December, Society for Underwater Technology, London, 35–40.

Rogers, S. M. (1987). Artificial seaweed for erosion control. *Shore and Beach*, American Shore and Beach Preservation Society, January, 19–29.

Sager, G. and Sammler, R. (1968). *Atlas der Gezeitenströme für die Nordsee, den Kanal und die Irisch See* (Zweite Verbenerte Auflage), Seehydrographischer Dienst der DDR, Rostock.

Saito, E., Sata, S. and Shibayama, T. (1990). Local scour around a large circular cylinder due to wave action, Abstract 95, *Proc. 22nd Int. Conf. Coastal Engng*, Delft.

Saito, E. and Shibayama, T. (1992). Local scour around a large cylinder on the uniform bottom slope due to waves and currents, *Proc. 23rd Int. Conf. Coastal Engng*, Venice, 2799–2810.

Sakai, T., Hatanaka, K. and Mase, H. (1992). Wave-induced stresses in sea bed and its momentary liquefaction,. *Proc. Am. Soc. Civ. Engrs, J. Waterways, Port, Coastal and Ocean Engng*, **118**, (WW2), 202–206.

Sawaragi, T. (1992). Detached breakwaters, in: *Design and Reliability of Coastal Structures*, Proceedings of the Short Course attached to 23rd ICCE, Venice, October, 351–372.

Schwind, R. G. (1962). *The 3-dimensional boundary layer near a strut.* Gas Turbine Lab., Report 67, MIT.

Shen, H. W., Schneider, V. R. and Karaki, S. S. (1966). *Mechanisms of local scour*, Engng Res. Center, Colorado State University, Report CER66HWS22.

Silvester, R. and Hsu, J. R. C. (1997). *Coastal stabilisation*. World Scientific.

Slaattelid, O. H., Myrhaug, D. and Lambrakos, K. F. (1987). North Sea bottom boundary layer study for pipelines, *Proceedings 19th Annual Offshore Technology Conference*, Houston, 191–198.

Song, K. K., Kloth, H. L., Costello, C. R. and Liesesemer, S. V. (1979). Anti-scour method uses air lift idea, *Offshore*, Feb., 49–52.

Soulsby, R. L. (1983). The bottom boundary layer of shelf seas, in: *Physical Oceanography of Coastal and Shelf Seas*, ed. Johns, B., 189–266, Elsevier Oceanography Series, 35.

Soulsby, R. L. (1987). Calculating bottom orbital velocity beneath waves. *Coastal Engng*, **11**, 371–380.

Soulsby, R. L. (1990). Tidal-current boundary layers. in: *The Sea*, Le Mehaute B. and Hanes, D. M. (eds), Vol. 9A, Chapter 15, Wiley, Chichester.

Soulsby, R. L. (1993). The reference concentration for suspended sand in a steady current. *MAST-II G8 Coastal Morphodynamics Grenoble Overall Workshop*, 6–10 September , Abstracts-in-depth, Delft Hydraulics, Delft.

Soulsby, R. L. (1995). Bed shear-stresses due to combined waves and currents. Section 4.5 in *Advances in Coastal Morphodynamics*, Stive, M. J. F. *et al.* (eds), Delft Hydraulics, Delft.

Soulsby, R. L. (1997). *Dynamics of Marine Sands. A Manual for Practical Applications*. Thomas Telford, London.

Soulsby, R. L. and Whitehouse, R. J. S. (1997). Threshold of sediment motion in coastal environments. *Proc. Pacific Coasts and Ports '97 Conf.*, Christchurch, University of Canterbury, New Zealand, 144–154.

Soulsby, R. L., Hamm, L, Klopman, G, Myrhaug, D, Simons, R. R. and Thomas, G. P. (1993). Wave–current interaction within and outside the bottom boundary layer. *Coastal Engng*, **21**, 41–69.

Southgate, H. N., (1995). Prediction of wave breaking processes at the coastline, in: Rahman, M. (ed.), *Potential Flow of Fluids*, Vol. 6 of Advances in Fluid Mechanics, PCM Publications.

Staub, C. and Bijker, R. (1990). Dynamic numerical models for sandwaves and pipeline selfburial. *Proc. 22nd Int. Conf. Coastal Engng*, Delft, 2508–2521.

Stride, A. H. (ed.) (1984). *Offshore Tidal Sands. Processes and Deposits*, Chapman and Hall, London.

Stubbs, S. B. (1975). Sea bed foundation considerations for gravity structures, in: *Off-shore Structures*, ICE, London, 67–74.

Sumer, B. M., Christiansen, N. and Fredsøe, J. (1992a). Time scale of scour around a vertical pile. *Proc. 2nd International Offshore and Polar Engineering Conference*, ISOPE, San Francisco, Vol. 3, 308–315.

Sumer, B. M., Christiansen, N. and Fredsøe, J. (1993). Influence of cross section on wave scour around piles, *Proc. Am. Soc. Civ. Engrs, J. Waterway, Port, Coastal, and Ocean Engrg.*, **119**, (5), 477–495.

Sumer, B. M. and Fredsøe, J. (1990). Scour below pipeline in waves, *Proc. Am. Soc. Civ. Engrs, J. Waterway, Port, Coastal and Ocean Engng*, **116**, (3), 307–322.

Sumer, B. M. and Fredsøe, J. (1991). Onset of scour below a pipeline exposed to waves, *Proc. 1st Int. Offshore and Polar Engng. Conf.*, 290–295.

Sumer, B. M. and Fredsøe, J. (1993). A review of wave/current-induced scour around pipelines. *Proc. 23rd Conf. Coastal Engng*, 4–9 Oct. 1992, Venice, 2839–2852.

Sumer, B. M. and Fredsøe, J. (1994). Self-burial of pipelines at span shoulders, *Int. J. Offshore Polar Engng*, **4**, (1).

Sumer, B. M. and Fredsøe, J. (1997). Scour at the head of a vertical-wall breakwater. *Coastal Engrg*, **29**, 201–230.

Sumer, B. M., Fredsøe, J. and Christiansen, N. (1992b). Scour around vertical pile in waves, *Proc. Am. Soc. Civ. Engrs, J. Waterway, Port, Coastal, and Ocean Engng*, **118**, (1), 15–31.

Sumer, B. M., Fredsøe, J. and Laursen, T. S. (1991). Experimental studies on non-uniform oscillatory boundary layers. In: Proceedings EUROMECH 262, eds Soulsby, R. L. and Bettess, R., 71–77, Balkema, Rotterdam.

Sumer, B. M., Jensen, H. R, Mao, Y. and Fredsøe, J. (1988a). Effect of lee-wake on scour below pipeline in current. *J. Waterway, Port, Coastal, and Ocean Engng., Am. Soc. Civ. Engrs*, **114**, (5), 599–614.

Sumer, B. M., Mao, Y. and Fredsøe, J. (1988b). Interaction between vibrating pipe and erodible bed. *J. Waterway, Port, Coastal, and Ocean Engng, Am. Soc. Civ. Engng*, **114**, (1), 81–92.

Sweeney, M., Webb, R. M. and Wilkinson, R. H. (1988). Scour around jackup rig footings, Paper OTC 5764, *Offshore Technology Conference*, May, 171–180.

Sybert, J. H. (1963).Foundation scour and remedial measures for off-shore platforms. Paper SPE 485, *1st Conf. Drilling and Rock Mechanics*, Austin, Texas.

Tesaker, E. (1980). *Underwater investigations of scour and scour protection, Proc. One day Society for Underwater Technology seminar on scour prevention techniques around offshore structures*, 55–66, 16 December.

Torsethaugen, K. (1975). *Lokalerosjon ved store konstruksjoner. Modellforsøk.* SINTEF Report STF60 A75055, Norwegian Hydrotechnical Laboratory, Trondheim (in Norwegian).

Toue, T., Katsui, H. and Nadaoka, K. (1992). Mechanism of sediment transport around a large circular cylinder, *Proc. 23rd Int. Conf. Coastal Engng.*, 2867–2878, Venice.

US Army Corps of Engineers (1986). *Preliminary Data Summary FRF* (Duck, NC), September 1986. Report prepared for Office, Chief of Engineers, US Army by Field Research Facility, Coastal Engineering Center (unpublished).

Ushijima, S. (1995). Numerical prediction of local scour with 3D body-fitted coordinates. Volume 3 of *Proceedings HYDRA 2000*, XXVIth IAHR Congress, London (ed. A. J. Grass), 287–292, Thomas Telford.

Van Beek, F. A. and Wind, H. G. (1990). Numerical modelling of erosion and sedimentation around offshore pipelines, *Coastal Engng*, **14**, 107–128.

Van Dijk, R. (1980). *Experience of scour in the southern North Sea, Proc. one day Society for Underwater Technology seminar on scour prevention techniques around offshore structures*, 3–10, 16 December.

Verhey, H. J. (1983). The stability of bottom and banks subjected to the velocities in the propeller jet behind ships. *8th Int. Harbour Congr.*, Antwerp, 1983, June, 13–17. Also Delft Hydraulics Publication No. 303.

Verley, R. L. P., Moshagen, B. H., Moholdt, N. C. and Nygaard, I. (1994). Trawl forces on free-spanning pipelines. *Int. J. Offshore and Polar Engng*, **2**, 24–31.

Vitall, N., Kothyari, U. C. and Haghighat, M. (1994). Clear-water scour around bridge pier group, *Proc. Am. Soc. Civ. Engrs, J. Hydraulic Engng*, **120**, (11), 1309–1318.

Wagstaff, M. J. (1993). *Effective modelling of scour around bottom-founded, sub-surface structures*, unpublished MPhil thesis, University of Strathclyde.

Wallace, H. M. and Chesher, T. J. (1994). Validation of coastal morphodynamic models with field data. *Proceedings of Coastal Dynamics '94*, Barcelona, 114–128.

Waters, C. B. (1994). The HR Wallingford scour monitoring system. *Proc. 100th Anniversary Bridge Conf.*, University of Cardiff, 1994, Oct.

Watson, T. N. (1979). Scour in the North Sea, *J. Petroleum Technology*, **26**, (3).

Whitehouse, R. J. S. (1993). Combined flow sand transport: field measurements. *Proceedings 23rd International Conference on Coastal Engineering*, Venice, Vol. 3, 2542–2555.

Whitehouse, R. J. S. (1995). The transport of sandy sediments over sloping beds. Section 2.9 in *Advances in coastal morphodynamics*, (eds M. J. F. Stive *et al.*), Delft Hydraulics, Delft.

Whitehouse, R. J. S., Owen, M. W. and Stevenson, E. C. (1997). *Sediment transport measurements at Boscombe Pier, Pool Bay. Data Report*. HR Wallingford Report TR 27.

Wilkinson, R. H., Palmer, A. C., Ells, J. W., Seymour, E. and Sanderson, N. (1988). Stability of pipelines in trenches. *Proceedings of the Offshore Oil and Gas Pipeline Technology Seminar*, Stavanger.

Wilson, N. D. and Abel, W. (1973). Seafloor scour protection for a semi-submersible drilling rig on the Nova Scotian shelf, Paper OTC 1891, *Offshore Technology Conference*, Dallas.

Wimpey Offshore (1990). *Sea bed stability of debris*. Department of Energy, Offshore Technology Report OTH 89 313, HMSO.

Yanmaz, A. M. and Altinbilek, H. D. (1991). Study of time-dependent local scour around bridge piers, *Proc. Am. Soc. Civ. Engrs, J. Hydraul. Engng*, **117**, (10), 1247–1268.

Zen, K. and Yamakazi, H. (1993). *Wave-induced liquefaction in a permeable sea bed*. Report of the Port and Harbour Research Insitute, Japan, Vol. 6, 155–192.

Appendix 1: Pipeline scour model*

The following is based on the results from an investigation into the numerical modelling of local scour under pipelines completed by Dr B. Brørs during his EC (HCMP) funded study visit to HR Wallingford (1993–1994). This research helped to assess the feasibility of using such numerical models as tools for engineering applications (Chapter 4). A full description of the model can be found in Brørs (1997).

A1.1. GENERAL APPROACH

The approach used in numerical scour prediction is as follows (Figure 44):

- Run a flow model: in a steady flow case the flow model is run until a steady solution obtained.
- Run a bed change model: the change in bed levels throughout the domain is calculated by enforcing continuity of bed sediment, using as input data bed shear stresses, turbulent viscosities, sediment concentrations etc. produced by the flow model.
- Adapt the grid to the new geometry and repeat the procedure.

The most usual procedure for the bed change calculation is first to find the rate of change of bed level over the whole bed,

*Text provided by Dr B. Brørs, SINTEF Civil and Environmental Engineering, Trondheim, Norway.

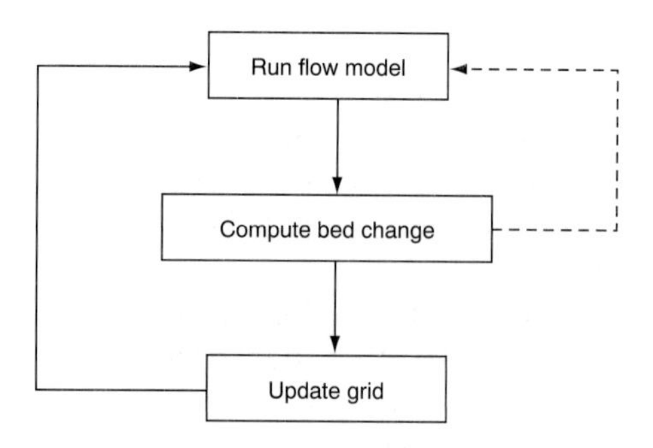

Figure 44. Procedure for numerical scour prediction

identify the local maximum change rate, and then choose a *morphological* time-step so large as to give a predetermined maximum allowed scour depth increase, say 10% of the present depth of the scour hole. After a grid update, the flow model should be run sufficiently long for transients to die out, say one to two times D/u_0 where D is the diameter of the structure or length of the bed feature and u_0 is the free stream flow velocity.

AI.2. FLOW MODEL

The equations to be solved are the Reynolds-averaged equations for fluid continuity (Equation (55)), fluid momentum (the Navier–Stokes equations) (Equation (56)), and possibly concentration c of suspended sediment (Equation (57)):

$$\frac{\partial u_i}{\partial x_i} = 0 \tag{55}$$

$$\frac{\partial u_i}{\partial t} + u_j \frac{\partial u_i}{\partial x_j} = -\frac{1}{\rho_f} \frac{\partial p}{\partial x_i} - \frac{\partial \overline{u_i' u_j'}}{\partial x_j} + \frac{\rho_s - \rho_f}{\rho_f} c\, g\, \delta_{i3} \tag{56}$$

$$\frac{\partial c}{\partial t} + (u_j - w_s \delta_{i3}) \frac{\partial c}{\partial x_j} = -\frac{\partial \overline{c' u_j'}}{\partial x_j} \tag{57}$$

The equations are expressed for a Cartesian coordinate system with the x_3 axis directed vertically upwards. The density of the suspension is expressed as $\rho = \rho_f + (\rho_s - \rho_f)c$ where subscripts s and f denote sediment and fluid respectively. Sediment-induced vertical density forces are included in the momentum Equation (56) by the last term, and the effect of a constant sediment settling velocity w_s (positive in the downward direction) on the sediment transport is included on the left-hand side of Equation (57).

The turbulent fluxes of momentum and sediment are unknown quantities. They could be found by solving their transport equations (a dynamic Reynolds stress turbulence model) but a more usual approach is to express the fluxes in terms of a turbulent viscosity ν_t and mean flow quantities using the Boussinesq approximation:

$$-\overline{u_i u_j} = \nu_t \left(\frac{\partial u_i}{\partial x_j} + \frac{\partial u_j}{\partial x_i} \right) - \frac{2}{3} \delta_{ij} k, \quad k = \frac{1}{2} \overline{u_k u_k} \; \overline{u'_k u'_k} \tag{58}$$

$$-\overline{c u'_j} = \frac{\nu_t}{\sigma_c} \frac{\partial c}{\partial x_j} \tag{59}$$

A turbulence model is still necessary in order to provide a value for the turbulent viscosity ν_t throughout the flow field. It can be a zero-equation (algebraic), one-equation (k) or two-equation k–ε model. The standard k–ε model equations are:

$$\frac{\partial k}{\partial t} + u_j \frac{\partial k}{\partial x_j} = \frac{\partial}{\partial x_j} \left(\frac{\nu_t}{\sigma_k} \frac{\partial k}{\partial x_j} \right) + P + G - \varepsilon \tag{60}$$

$$\begin{aligned} \frac{\partial \varepsilon}{\partial t} + u_j \frac{\partial \varepsilon}{\partial x_j} = & \frac{\partial}{\partial x_j} \left(\frac{\nu_t}{\sigma_\varepsilon} \frac{\partial \varepsilon}{\partial x_j} \right) \\ & + [C_{\varepsilon 1} P - C_{\varepsilon 2} \varepsilon + C_{\varepsilon 3} \max\,(0, G)] \frac{\varepsilon}{k} \end{aligned} \tag{61}$$

The turbulent viscosity is defined as $\nu_t = C_\mu k^2/\varepsilon$ with k and ε given from Equations (60) and (61). P denotes the production rate of turbulent kinetic energy k by velocity shear and G denotes the production rate of k by gravity, defined as

$$P = \nu_t \frac{\partial u_i}{\partial x_j} \left(\frac{\partial u_j}{\partial x_i} + \frac{\partial u_i}{\partial x_j} \right), \quad G = \frac{\rho_s - \rho_f}{\rho_f} g \frac{\nu_t}{\sigma_c} \frac{\partial c}{\partial x_j} \tag{62}$$

A drawback of zero-equation models is that they require the turbulent viscosity to be specified throughout the flow field. This can be difficult to do in a complex domain with possible recirculation zones, and the empirical relations at hand may not be valid. The same goes for k models where the turbulent viscosity is specified as $\nu_t = C_\mu^{1/4} k^{1/2} / l$. Here k is found by solving Equation (60) but the turbulent length scale l must be specified. However, this should be easier than specifying ν_t. The k–ε model is more general in that the viscosity is expressed in terms of k and ε and no empirical relations have to be produced, but the penalty is two extra equations to solve and possibly a less robust model requiring shorter time-steps. Hoffmans and Booij (1993) use a zero-equation turbulence model in scour calculations, whereas Van Beek and Wind (1990), Olsen and Melaaen (1993), and Brørs (1997) use a k–ε model. A survey of turbulence models can be found in Rodi (1984).

When it comes to discretisation of the equations, both finite difference (or finite volume) and finite element methods have been used in scour calculations. It is an advantage to use a model with a grid that can be easily fitted to a changing bed and organised so as to give higher spatial resolution where it is needed and less where it is not.

AI.3. BED CHANGE MODEL

The requirement of mass balance of bed sediment implies that the rate of change of the bed level can be expressed as

$$\frac{\partial h}{\partial t} = \frac{1}{1-n}\left[-\left(\frac{\partial q_b}{\partial x} + \frac{\partial q_b}{\partial y}\right) + D(x, y) - E(x, y)\right] \tag{63}$$

Considering a unit area of the bed, the equation states that the rate of change in bed level h equals the change in bedload flux q_b across the unit area plus the rate $D(x, y)$ at which sediment volume is deposited by the flow minus the rate $E(x, y)$ at which it is entrained by the flow. It is necessary to multiply with the factor containing the bed porosity n in order to translate from sediment volume flux to actual bed volume including pores. The bedload transport rate is usually expressed in terms of a bedload formula such as

$$q_b^* = 12\theta^{1/2}(\theta - \theta_{cr})[g(s-1)d^3]^{1/2} \tag{64a}$$

$$q_b = q_b^* - C|q_b^*|\frac{\partial h}{\partial x} \tag{64b}$$

where q_b^* is the horizontal bedload transport rate (Soulsby, 1993) and the second term on the right hand side of equation (64) is a slope correction term (Chesher *et al.*, 1993). Laboratory measurements (not local scour but sediment transport in general) seems to indicate that C is in the range 1·5–2·3 (Whitehouse, R. J. S., personal communication, 1994), $C = 1{\cdot}5$ is used here. The critical Shields parameter is adjusted so as to give a higher threshold for sediment motion up a slope and lower threshold for motion down a slope compared to the horizontal bed threshold θ_{cr}, using the formula $\theta_{\beta cr} = \theta_{cr} \sin(\phi_i \pm \beta)/\sin\phi_i$ where β is the slope angle and ϕ is the internal friction angle of the sediment (based on Equation (33) in the main text). The deposition rate is equal to the sediment settling velocity times the near-the-bed concentration, $D(x, y) = w_s c_0$, and the erosion rate is expressed in terms of the local near-the-bed turbulent viscosity and concentration gradient, $E(x, y) \propto \nu_t \, dc/dz$.

Equation (63) may look simple, but because of the non-linear bedload formula (Equations (64a), (64b)), problems with numerical instability in bed update schemes are often encountered.

AI.4. NOTATION IN APPENDIX I

- c sediment concentration (Reynolds-average)
- $\overline{c'u'_j}$ turbulent scalar flux (j component)
- C_μ constant, here equal to 0·09
- d sediment grain diameter (average)
- D sediment deposition rate (volumetric)
- E sediment erosion rate (volumetric)
- g gravitational acceleration
- h bed level
- k turbulent kinetic energy
- l turbulent length scale
- p dynamical pressure (Reynolds-average)
- q_b sediment bedload transport rate (volumetric)
- q_b^* sediment bedload transport rate, horizontal bed (volumetric)
- s relative density, $s = \varrho_s/\varrho_f$
- u_i flow velocity in the x_i direction (Reynolds-average)

u_* friction (or shear) velocity, $u_* = (\tau/\rho)^{1/2}$
$\overline{u_i'u_j'}$ turbulent momentum flux or Reynolds stress (i, j component)
w_s sediment settling velocity
x_i Cartesian space coordinate (with $i = 3$ denoting the vertical direction)
δ_{ij} Kronecker delta, $\delta_{ij} = 1$ when $i = j$; else $\delta_{ij} = 0$
ε rate of dissipation of turbulent kinetic energy
θ Shields parameter, $\theta = u_*^2/g(s-1)d$
$\theta_{\beta cr}$ critical Shields parameter (threshold for bedload transport, sloping bed)
θ_{cr0} critical Shields parameter (threshold for bedload transport, horizontal bed)
ν molecular diffusivity
ν_t turbulent diffusivity
ϱ local density (Reynolds-average)
ϱ_f fluid density
ϱ_s sediment density
σ_c turbulent Prandtl–Schmidt number for c
σ_k turbulent Prandtl–Schmidt number for k
σ_ε turbulent Prandtl–Schmidt number for ε
ϕ_i sediment angle of repose

AI.5. REFERENCES

See main reference list.

Appendix 2: Calculation methods for hydrodynamics and sediment parameters

Unless otherwise specified in the text, the following methods have been used to derive quantities referred to in this book. References for these methods are indicated. For convenience the SI units for quantities are given, but, of course, any consistent set of units can be used. Notation is as used in the main text.

A2.1. CURRENT RELATED BED SHEAR STRESS (SOULSBY, 1997)

The grain related bed shear stress due to steady current τ_c ($\mathrm{N\,m^{-2}}$) is calculated assuming:

$$\tau_c = \rho C_D \bar{U}^2 \tag{65}$$

where ρ is the fluid density ($\mathrm{kg\,m^{-3}}$) and C_D is the dimensionless drag coefficient for the depth-averaged current speed $\bar{U}$ ($\mathrm{ms^{-1}}$):

$$C_D = \left[\frac{0{\cdot}40}{\ln(h/z_0) - 1}\right]^2 \tag{66}$$

in which h is the mean water depth (m) and z_0 the hydraulic roughness length (m) relating to the bed sediment. It has been assumed that:

$$z_0 = \frac{k_s}{30} = \frac{2{\cdot}5 d_{50}}{30} \tag{67}$$

where d_{50} is the median diameter (m) of the sediment. In this method, as in those below, a single representative grain size other than d_{50} can be used if appropriate.

A2.2. WAVE RELATED BED SHEAR STRESS (SOULSBY, 1997)

The wave related bed shear stress τ_w (N m^{-2}) is calculated assuming:

$$\tau_w = 0{\cdot}5\rho f_w U_w^2 \tag{68}$$

where f_w is the dimensionless wave friction factor and U_w is the amplitude of the bottom orbital velocity due to the wave motion (m s^{-1}). The expression for f_w in rough turbulent flow is:

$$f_w = 1{\cdot}39(A/z_0)^{-0{\cdot}52} \tag{69}$$

where A is the amplitude of the orbital wave motion at the bed and z_0 is calculated for the bed grains using Equation (67). The expression for A is

$$A = \frac{U_w T_p}{2\pi} \tag{70}$$

where T_p is the wave period(s) at the peak of the surface elevation spectrum.

A2.3. COMBINED WAVE AND CURRENT SHEAR STRESS (SOULSBY, 1995 AND 1997)

The following approach for calculating the maximum bed shear stress τ_{max} (N m^{-2}) due to the interaction of waves and currents has been adopted. Firstly the mean shear stress τ_m (N m^{-2}) due to combined waves and currents is calculated:

$$\frac{\tau_m}{\tau_c} = 1 + 1{\cdot}2\left(\frac{\tau_w}{\tau_c + \tau_w}\right)^{3{\cdot}2} \tag{71}$$

where τ_w is calculated from Equation (68) and τ_c from Equation (65). This expression accounts for the non-linearities introduced when waves and currents interact and has been calibrated

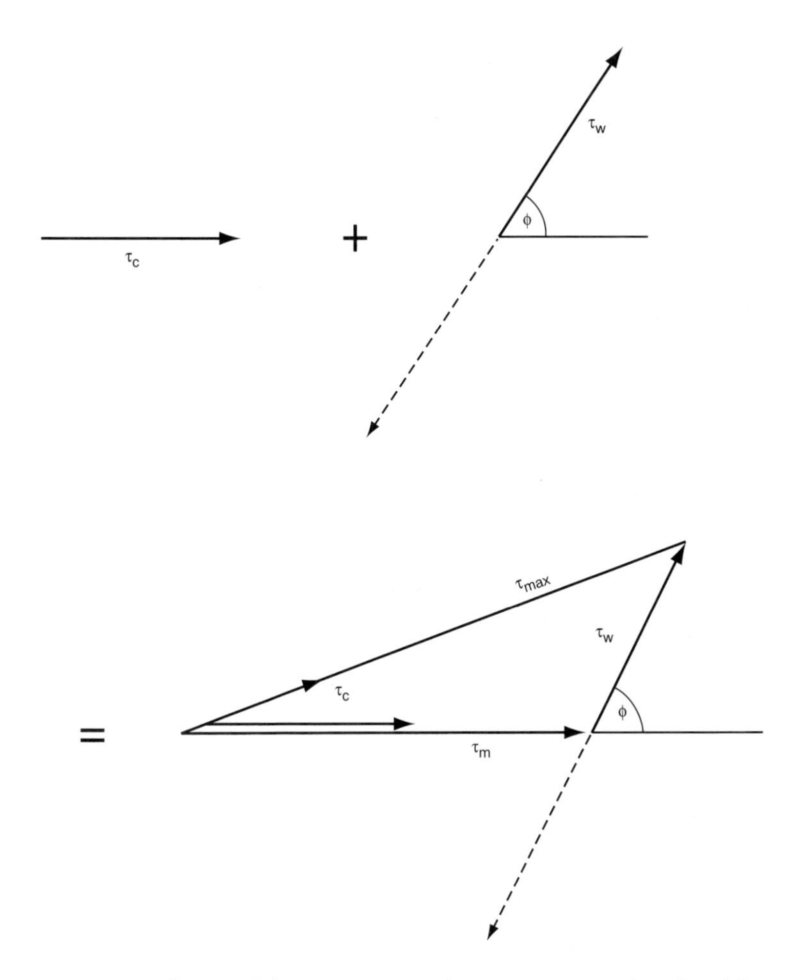

Figure 45. Definition of shear stress vectors for wave, current and combined shear stresses

against laboratory and field data (DATA2 formula of Soulsby, 1995 and 1997).

The maximum shear stress is calculated by vector addition of τ_m and τ_w to give the magnitude of the shear stress vector (Ockenden and Soulsby, 1994, Figure 45):

$$\tau_{max} = [(\tau_m + \tau_w \cos\phi)^2 + (\tau_w \sin\phi)^2]^{0.5} \quad (72)$$

where ϕ is the angle (°) between the wave and current shear stresses, or velocities to a reasonable approximation. When it is known, the angle between waves and currents should be used in Equation (72), but an *average* value of ϕ equal to 45° can be assumed in place of a known value.

A2.4. NON-DIMENSIONAL BED SHEAR STRESS

The values of bed shear stress τ_o are non-dimensionalised using the Shields parameter $\theta_{2 \cdot 5}$, where the subscript 2·5 refers to the grain size multiplier for k_s in Equation (67):

$$\theta_{2 \cdot 5} = \frac{\tau}{(\rho_s - \rho) g d_{50}} \tag{73}$$

Here ρ_s is the grain mineral density ($\mathrm{kg\,m^{-3}}$) and g the acceleration due to gravity ($9 \cdot 81\,\mathrm{m\,s^{-2}}$).

A2.5. THRESHOLD FOR MOTION OF COHESIONLESS SEDIMENT (SOULSBY, 1997)

Calculate the dimensionless grain size:

$$D_* = d_{50} \left(\frac{(s-1)g}{\nu^2} \right)^{1/3} \tag{74}$$

where $s = \rho_s/\rho$ and ν is the kinematic viscosity of the fluid ($\mathrm{m^2\,s^{-1}}$).

An algebraic representation for this form of the Shields curve has been derived:

$$\theta_{2 \cdot 5,\mathrm{cr}} = \frac{0 \cdot 24}{D_*} + 0 \cdot 055 \left[1 - \exp(-0 \cdot 02 D_*) \right] \tag{75a}$$

For $D_* < 10$ (fine sand) an alternative expression is given by Soulsby (1997) and Soulsby and Whitehouse (1997) which fits well to data:

$$\theta_{2 \cdot 5,\mathrm{cr}} = \frac{0 \cdot 30}{1 + 1 \cdot 2 D_*} + 0 \cdot 055 \left[1 - \exp(-0 \cdot 020 D_*) \right] \tag{75b}$$

This expression is identical to Shield's curve for D_* greater than about 10 (Figure 46).

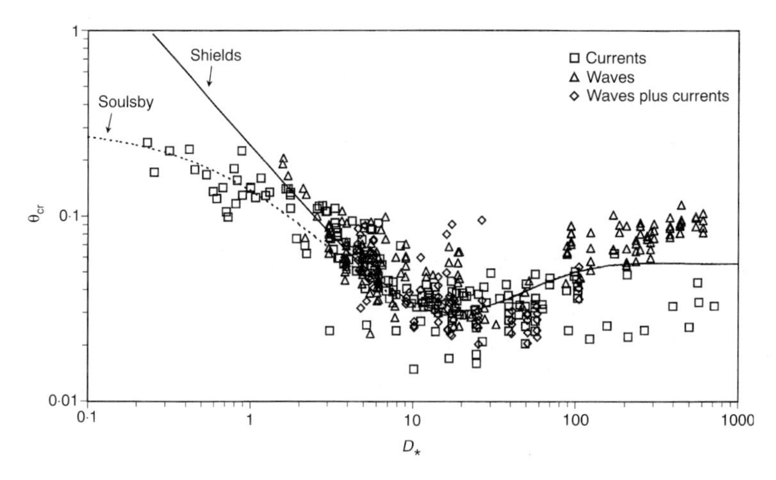

Figure 46. Threshold of motion of sediments beneath waves and/or currents (reproduced from Soulsby, 1997).

A2.6. SETTLING VELOCITY OF COHESIONLESS SEDIMENT (SOULSBY, 1997)

The settling velocity w_s ($\mathrm{m\,s^{-1}}$) of individual grains can be calculated from the following formula which has been optimised against data:

$$w_s = \frac{\nu}{d_{50}}\left[\left(10{\cdot}36^2 + 1{\cdot}049 D_*^3\right)^{1/2} - 10{\cdot}36\right] \tag{76}$$

A2.7. THRESHOLD FOR COHESIVE MIXTURES OF MUD AND SAND (MITCHENER *ET AL.*, 1996)

The variation of critical shear stress for erosion with bulk density of the sediment ρ_b ($\mathrm{kg\,m^{-3}}$) is given by

$$\tau_{cr} = 0{\cdot}015(\rho_b - 1000)^{0{\cdot}73} \tag{77}$$

The dimensional constant gives τ_{cr} in $\mathrm{Nm^{-2}}$ and it is usually the case that cohesive sediment erosion stress is used in a dimensional form rather than as Equation (73). Typical values of z_0 for calculating τ_0 are (mud) 0·2 mm, (mud/sand) 0·7 and (silt/sand) 0·05 mm (Soulsby, 1997).

A2.8. REFERENCES

See main reference list.

Index